Active Filter Design

Active Filter Design

Allan Waters

Nottingham Polytechnic

Macmillan New Electronics
Introductions to Advanced Topics

McGraw-Hill, Inc.

New York   St. Louis   San Francisco   Bogotá   Caracas
Mexico   Montreal   San Juan   São Paulo   Toronto

1 2 3 4 5 6 7 8 9 0   DOC/DOC   9 7 6 5 4 3 2 1

ISBN 0-07-068453-7

First published 1991 by MACMILLAN EDUCATION LTD.
Houndmills, Basingstoke, Hampshire RG21 2XS
and London.

Printed and bound by R. R. Donnelley & Sons Company.

Dedicated to Kay, Duncan and Catherine

Contents

Series Editor's Foreword

The rapid development of electronics and its engineering applications ensures that new topics are always competing for a place in university and polytechnic courses. But it is often difficult for lecturers to find suitable books for recommendation to students, particularly when a topic is covered by a short lecture module, or as an 'option'.

Macmillan New Electronics offers introductions to advanced topics. The level is generally that of second and subsequent years of undergraduate courses in electronic and electrical engineering, computer science and physics. Some of the authors will paint with a broad brush; others will concentrate on a narrower topic, and cover it in greater detail. But in all cases the titles in the Series will provide a sound basis for further reading of the specialist literature, and an up-to-date appreciation of practical applications and likely trends.

The level, scope and approach of the Series should also appeal to practising engineers and scientists encountering an area of electronics for the first time, or needing a rapid and authoritative update.

Paul A. Lynn

Preface

My principal objective in writing this book is to present the basic concepts of active filtering in a way which, I hope, will be readily understood by undergraduate and BTEC HND students, and may also be found acceptable by postgraduate students who are unfamiliar with the subject.

Once the basic concepts have been assimilated, it is possible to comprehend the extensive published literature which covers all aspects of active filters. Indeed, I am indebted to many sources for much of the information in the book, not least the references given at the end of the book which have influenced me considerably.

The structure of the book is based on analysis, followed by synthesis; this approach has been favourably received by a succession of engineering students. While not claiming to treat all topics exhaustively, my aim has been to introduce, concisely, each underlying principle involved in the comprehensive design of a chosen filter. The books and articles in the Bibliography will provide more than adequate further information for those readers requiring a more descriptive and analytically detailed approach to the circuits.

It is my experience with undergraduates that achieving a successful design is paramount either for part of a complex project, or more mundanely for boosting confidence. The finer points and subtleties of the design procedure may then be studied later.

Chapter 1 introduces the concept of filtering and miscellaneous details, such as component quality, scaling and so on. The second chapter introduces the reader to the Butterworth and Chebyshev responses; this is followed, in chapter 3, by a brief overview of the operational amplifier.

Chapter 4 brings together the ideas in the preceding chapters and applies them to the popular VCVS and MFB circuits, along with a brief review of higher-order circuits.

Chapter 5 introduces the biquadratic circuits and includes a résumé of the switched-capacitor filter (SCF) and the all-resistor biquad.

Chapter 6 is devoted to an appreciation of passive circuits and concentrates on the synthesis technique. It also includes a brief discussion of the principle of frequency transformation.

Chapter 7 uses the passive circuit realisations of chapter 6 and introduces the reader to the gyrator, GIC and FDNR.

The final chapter shows how circuits behave for changes in component values – the so-called sensitivity analysis.

The book concludes with a short appendix outlining a computer aided design (CAD) technique which it is felt can considerably assist filter design.

Each chapter contains worked examples. In addition, there are problems, with answers, included at the end of most chapters.

Finally, I should like to record here my appreciation to those who have supported me in the writing of this book. I am indebted to my family for their continued support and encouragement, Barbara Sandford who produced the typescript, and Lynnda Aucott who drew the diagrams.

Allan Waters

Active Filter Design

1 Introduction

1.1 Filter circuit concepts

A filter is an electrical network which is designed specifically to modify, in a prescribed way, an electrical signal applied to its input terminals. An elementary example would be a signal having only low-frequency and high-frequency components. A *low-pass* filter would permit only the low frequencies to be transmitted, the high frequencies being suppressed or attenuated. The converse is true for passing the high frequencies while suppressing the low frequencies by means of a *high-pass* filter.

Other filters may be designed to pass a range of frequencies about a given selected frequency, rejecting frequencies outside the selected band. Conversely, a filter may be designed to reject a narrow band of frequencies while accepting all other frequencies outside the selected band. It may be said that electrical and electronic filter networks have become an indispensable part of electronic and modern communication systems.

The fundamental principles of electrical wave filters were outlined by Wagner in Germany and Campbell in the USA around the year 1915. Since those early beginnings, the *state-of-the-art* has encompassed many new ideas and techniques associated with modern network theory and feedback analysis. Examples include the work of Bode and Black on the stability of amplifiers and that of Darlington in the application of modern network theory to electrical filters.

The choice of a particular type of filter from the large range of filters must be influenced by several factors, some of which are: complexity, ease of tuning, economics, compatability with existing circuitry and whether or not power supplies may be required.

By referring to figure 1.1 it can be seen that active filter realisations are more advantageous at lower frequencies than is the case for passive types, although the reverse is true at higher frequencies. Furthermore, at frequencies close to or within the microwave range, only waveguide or coaxial cable type filters are possible.

Mention must be made of *digital filtering techniques* since such filters are becoming widely used and of increasing importance. Simply stated, these circuits perform the filtering task using analogue and digital elements including A/D converters, D/A converters, shift registers, multipliers and multiplexers, and were indeed first implemented using digital computers. Digital filters far exceed the analogue circuits where high-order filters and multiplexing are required, since

1

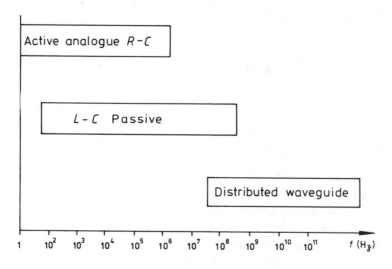

Figure 1.1   Comparison of frequency range of filters

tuning may be simply achieved by adjusting the coefficients of a mathematical algorithm. A disadvantage is that the filtering of high frequencies is limited by the time delays within the circuitry, but improvements are being made with the introduction of faster semiconductor materials such as gallium arsenide.

Other types of filters are the *piezo-crystal* type which presents very stable resonant frequencies, very high Q-factors and small power losses; *the surface acoustic wave* SAW type used in the majority of TV receivers and, finally, *mechanical* filters, a popular type being based on the principle of magneto-striction. This book will concentrate almost wholly on the $R-C$ types with a brief mention of the passive types and their op-amp, gyrator equivalents.

1.2 Advantages of active over passive circuits

Passive filters are constructed from combinations of resistance and capacitance or inductance and capacitance and can be designed to cover a wide band of frequencies (typically 10 Hz to 500 MHz). They have the advantage over active circuits in that they do not require an external power supply and that they have low sensitivities to component variation. At low frequencies where cost and size are important considerations, there arises the problem of large coils having low Q factors. The coils usually require efficient magnetic cores and special winding techniques to enable a reasonable level of quality to be achieved, all of which results in increased costs.

Active filters are constructed from resistors, capacitors and an active source which is usually an operational amplifier. The main advantages over the passive types are that expensive and bulky coils are eliminated, circuit gain may be realised, and high input and low output impedances give the amplifier good isolating properties which are useful when cascading networks to produce higher-order filters.

The main disadvantages are that a power supply is required, that above about 50 kHz the op-amp has a considerably reduced gain (although more expensive devices are available to the designer for frequencies up to 500 kHz) and that the circuit sensitivities to component changes are worse than the passive counter-parts. At high frequencies, coils are of considerably improved quality, cheaper and physically much smaller, thereby making the passive *L-C* circuit a more viable proposition at such frequencies.

1.3 Component quality

Clearly it is of paramount importance to select good-quality components having a high stability over the dynamic working range. Care must therefore be taken in the choice of op-amps, resistors and capacitors when constructing active filter circuits. The op-amp is considered in chapter 3 where the important user design data is *gain–bandwidth product, slew rate* and whether internal or external *compensation* is to be employed.

The types of resistors used are: *carbon-composition, carbon-film* and *metal-film.*

Carbon-composition resistors are probably the most widely used where resistance values are maintained around ±10 per cent by the manufacturing process, although values within ±5 per cent are obtained by using an automatic sorting technique.

Carbon-film resistors are more stable than the composition type and have very good temperature coefficients of resistance with good ageing characteristics. Typical tolerances are ±5 per cent which makes this resistor well suited to general-purpose applications.

Metal-film resistors are the most widely used for precision requirements where tolerances within the range ±1 per cent to ±0.1 per cent are available. They exhibit better temperature stability and ageing characteristics than the carbon film type.

Choice of type of capacitor may be made from film (polyester types), mica, ceramic or electrolytic depending on factors such as frequency range, temperature stability and cost.

Film types should be considered for filters operating below a few hundred kiloherz and up to 125°C. They are normally available with standard tolerances of ±1 per cent, ±10 per cent and ±20 per cent.

Mica capacitors are more expensive than the film types but possess far superior properties such as temperature coefficient, where a few parts in a million may be achieved. These capacitors may be used well into the VHF region.

Ceramic capacitors have greater dielectric constants than the film and mica types and may be constructed to provide positive or negative temperature co-efficients. This property makes these devices particularly useful for temperature compensation of active and passive circuits. They also exhibit minimum parasitic inductance and power dissipation over a range from low audio to high radio frequencies which make them attractive at radio frequencies. *Aluminium electro-lytic capacitors* are usually employed as by-pass devices or for non-precision timing applications at low frequencies and are in general unsuitable for active or passive circuits. The *tantalum* form, however, may be used at low frequencies where low selectivity filters are required.

1.4 Filter responses

There are five basic forms of filter-response and these are shown in figure 1.2 where a second-order function for $H(j\omega)$ is assumed.

From a consideration of the gain and phase responses in figure 1.2, it can be seen that the denominator expressions for $H(j\omega)$ are similar in form but have differences in order to accommodate the different design requirements. For example, the low- and high-pass filters must involve not only the pass and stop-band information, but also the type of response – whether Butterworth or Chebyshev. The latter is controlled by the coefficients b_1 and b_0. The band-pass, band-reject and all-pass circuits must involve design data based on Q-factor, bandwidth, centre frequency and centre frequency gain.

Finally, it should be noted that the total phase shift for the circuits is $180°$. This does not necessarily imply 0–$180°$, as will be seen in chapter 4.

1.5 Basic filter theory

A filter may be considered to be composed of lumped elements as shown in figure 1.3 where the input and output signals are continuous functions of time.

Using the Laplace operator s, where s is a complex variable, we may write the output/input relationship for the filter $H(s)$ in the form:

$$\frac{V_o}{V_i}(s) = H(s) \tag{1.1}$$

where $H(s)$ is called the *transfer function* of the filter.

We usually require to know how the filter responds to changing frequency so that equation (1.1) is modified by writing $s = j\omega$:

$$\frac{V_o}{V_i}(j\omega) = H(j\omega) \tag{1.2}$$

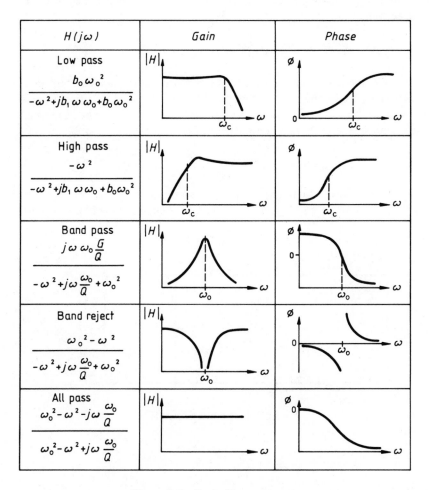

$H(j\omega)$	Gain	Phase
Low pass $\dfrac{b_0\omega_0^2}{-\omega^2+jb_1\omega\omega_0+b_0\omega_0^2}$		
High pass $\dfrac{-\omega^2}{-\omega^2+jb_1\omega\omega_0+b_0\omega_0^2}$		
Band pass $\dfrac{j\omega\omega_0\frac{G}{Q}}{-\omega^2+j\omega\frac{\omega_0}{Q}+\omega_0^2}$		
Band reject $\dfrac{\omega_0^2-\omega^2}{-\omega^2+j\omega\frac{\omega_0}{Q}+\omega_0^2}$		
All pass $\dfrac{\omega_0^2-\omega^2-j\omega\frac{\omega_0}{Q}}{\omega_0^2-\omega^2+j\omega\frac{\omega_0}{Q}}$		

Figure 1.2   Basic filter responses

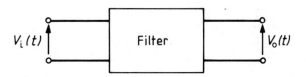

Figure 1.3   Two-port filter circuit

This may be re-written to provide the *magnitude* and *phase* components of the transfer function:

$$\frac{V_o}{V_i}(j\omega) = |H(j\omega)| \exp[j\phi(j\omega)] \tag{1.3}$$

$|H(j\omega)|$ is the *magnitude* and $\phi(j\omega)$ is the *phase* of the filter transfer function.

A knowledge of the transfer function enables us to anticipate the response of a particular filter design, and a simple *passive* circuit will now be used to introduce a number of important concepts.

Example 1.1

Obtain the transfer function, gain and phase responses for the circuit shown in figure 1.4, where R_S and R_L are the normalised source and load terminations.

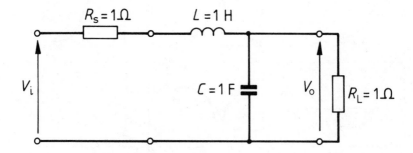

Figure 1.4   Simple passive filter

Elementary analysis yields the expression:

$$\frac{V_o}{V_i}(j\omega) = \frac{R_L}{R_L + R_S - \omega^2 LCR_L + j\omega(L + CR_LR_S)} \tag{1.4}$$

which becomes upon insertion of the component values:

$$\frac{V_o}{V_i}(j\omega) = H(j\omega) = \frac{1}{(2 - \omega^2) + j2\omega}$$

giving

$$|H(j\omega)| = \frac{1}{\sqrt{\{(2 - \omega^2)^2 + 4\omega^2\}}} \tag{1.5}$$

$$\phi(j\omega) = -\tan^{-1}\left(\frac{4\omega^2}{2 - \omega^2}\right)$$

Notice the circuit is that of a second-order filter since it is constructed using two energy storage elements.

In order to plot the gain and phase responses, it is necessary to select values for ω and to insert these values into the appropriate equations (1.5). Experience shows that usually three well-chosen values for ω are sufficient to enable reasonable response curves to be drawn when considering reasonably simple filter structures.

$$|H(j0)|_{\omega=0} = \frac{1}{2} \, ; \, \phi(j0)_{\omega=0} = 0°$$

$$|H(j\sqrt{2})|_{\omega=\sqrt{2}} = \frac{1}{2\sqrt{2}} \, ; \; \phi(j\sqrt{2})_{\omega=\sqrt{2}} = -\frac{\pi}{2}$$

$$|H(j\infty)|_{\omega=\infty} = 0 \, ; \; \phi(j\infty)_{\omega=\infty} = -\pi$$

Difficulty is often experienced in obtaining the phase characteristics of networks, and the method of obtaining such characteristics is best understood from a consideration of the denominator expression:

$$D = (2 - \omega^2) + j2\omega$$

It can be seen that for $\omega \leqslant \sqrt{2}$ then the real part of D is positive and that the angle increases from $0°$ to $90°$. For values of $\omega \geqslant \sqrt{2}$ then the real part of D becomes negative such that D lies in the second quadrant and the angle increases from $90°$ to $180°$. Since we are considering the denominator expression of $H(j\omega)$, the phase angle will be negative as shown in figure 1.5.

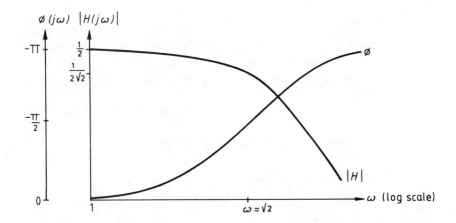

Figure 1.5 Response for example 1.1

The gain and phase responses for the circuit are shown in figure 1.5 plotted to a *logarithmic* frequency scale; a logarithmic frequency scale will be used throughout the book unless otherwise stated, since this allows compression of the frequency scale.

The voltage gain may be expressed in decibels by using the expression:

$$dB = 20\log_{10}|H(j\omega)| \tag{1.6}$$

Applying equation (1.6) to the voltage gain of the filter circuit under consideration gives the expression:

$$dB = 20\log_{10}\frac{1}{\sqrt{\{(2-\omega^2)^2 + 4\omega^2\}}} = -10\log_{10}\{(2-\omega^2)^2 + 4\omega^2\}$$

At $\omega = 0$, $dB = -10\log_{10}2^2 = -6$ dB

$\omega = \sqrt{2}$, $dB = -10\log_{10}8 = -9$ dB

$\omega \gg \sqrt{2}$, $dB = -10\log_{10}\omega^4 = -40\log_{10}\omega$

From the results it can be seen that the 3 dB *bandwidth* of the response occurs when $\omega = \sqrt{2}$ and for the condition that ω is very large then the gain *rolls off* at 40 dB per decade change of frequency. We may summarise the performance of any order n of filter having the *maximally flat* Butterworth characteristic shown in figure 1.5, as having a roll-off given by $20n$ dB/decade and having a phase shift of $n\pi/4$ at the 3 dB frequency with a final phase shift of $n\pi/2$ at frequencies far removed from the cut-off frequency.

1.6 Magnitude and frequency scaling

In example 1.1 the component values were conveniently chosen as unity (1 ohm, 1 Farad, 1 Henry). The process of referring components and also frequency to the value unity is referred to as *normalising* the circuit. Correctly applied, this *scaling* technique may significantly simplify the analysis and synthesis of electrical networks.

In example 1.1 the terminating resistors may have the value 600 ohms while the working bandwidth may extend to 1 kHz. Clearly the values of inductance and capacitance must change because of the influence of the actual design data.

1.6.1 Magnitude scaling factor (K_m)

Assume R_n, C_n, L_n are the *normalised* values of resistance, inductance and capacitance and that R_0, L_0, C_0 are the corresponding *scaled* values. Then clearly

$$R_0 = K_m R_n$$

The criterion is that the reactance values should remain unchanged after the application of magnitude scaling, thereby preserving the shape of the response characteristic.

$$\omega L_0 = K_m(\omega L_n), L_0 = K_m L_n$$

$$\frac{1}{\omega C_0} = \frac{K_m}{\omega C_n}, C_0 = \frac{C_n}{K_m}$$

1.6.2 Frequency scaling factor ($K_f = \omega_0/\omega_n$)

The criterion above must still hold for the reactance values after scaling has been applied when the frequency changes from ω_n to ω_0, again to preserve the shape of the response characteristic when moving to higher frequencies.

$$\omega_n L_n = \omega_0 L_0, L_0 = \left(\frac{\omega_n}{\omega_0}\right) L_n = \frac{L_n}{K_f}$$

$$\frac{1}{\omega_0 C_0} = \frac{1}{\omega_n C_n}, C_0 = \left(\frac{\omega_n}{\omega_0}\right) C_n = \frac{C_n}{K_f}$$

Consideration of figure 1.6 shows that the frequency scaling factor shifts each frequency from the normalised circuit to the scaled circuit *in the same proportion.*

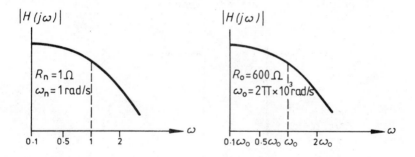

Figure 1.6   Frequency scaling

The two scaling factors may be combined as follows:

$$R_0 = K_m R_n$$

$$L_0 = \frac{K_m}{K_f} L_n \qquad\qquad (1.7)$$

$$C_0 = \frac{C_n}{K_m K_f}$$

Returning to Example 1.1, assume that the circuit is scaled for $R_L = R_S = 600\ \Omega$, $\omega_0 = 2\pi \times 10^3$ rad/s. Hence

$$L_0 = \frac{600 \times 1}{2\pi \times 10^3} = 0.095 \text{ H}$$

$$C_0 = \frac{1}{600 \times 2\pi \times 10^3} = 0.265 \ \mu F$$

1.7 Effect of complex signals

We have previously considered how the magnitude and phase responses of a filter circuit are affected by a change in the frequency of the input signal where the input was a *pure* sine wave. However, in many filter applications, the input signal is a complex waveform which is made up from an infinite number of sine waves; a typical input signal may be considered to be the so-called *square* form shown in figure 1.7. The Fourier series for such a waveform may be represented by the expression

$$f(t) = A \sin \omega_0 t + \frac{A}{3} \sin 3 \omega_0 t + \frac{A}{5} \sin 5 \omega_0 t + \ldots \tag{1.8}$$

where only the first few terms are of any signifiance.

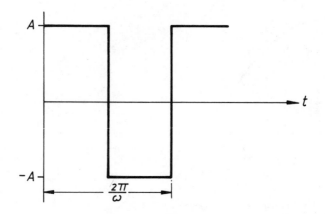

Figure 1.7 Square-wave input

The normal low-pass filter will be tuned to the fundamental component ω_0, all other components ($3\omega_0$, $5\omega_0$ etc.) being suppressed.

There is another type of filter, however, which is used to reproduce, ideally, at its output terminals, an exact replica of the input waveform such as that described by equation (1.8).

The problem arises when such a complex waveform is applied as the input signal to a filter. The fundamental along with all the significant harmonics must show the correct relative amplitude relationship at the output of the filter to maintain the original input signal. Furthermore, the various frequency components should not be displaced in time relative to one another. Filters designed around a constant delay for a given range of frequencies are called Bessel or Thomson filters and are not considered in this book.

Consider the response diagram shown in figure 1.5 and further assume that the phase characteristic is linear from dc to around the cut-off frequency (f_0). A sine-wave input signal is displaced in time at the output and may be determined by using the expression

$$\tau_p = \frac{\beta}{\omega} \tag{1.9}$$

where β is the phase shift in radians
ω is the input frequency in radians/second
τ_p is the phase delay.

For a linear phase response, the phase delay for each frequency component would be the same, which means that the output waveform would be equivalent to the input but displaced by a time τ.

Should the phase response be non-linear, then clearly the phase delay for each frequency would be different and output distortion would occur. Most filters have non-linear phase response characteristics, with the result that waveform distortion will occur for complex input signals.

Finally it should be mentioned that in amplitude modulated signals where the *carrier* is associated with *side-bands*, then the carrier, which is affected by the phase delay, could be delayed differently from the side-bands which are affected by what is called *group-delay*. This again could produce a distorted output signal.

2   Filter Types

2.1 Introduction

Modern filter design is based on the accurate selection of an appropriate input/output relationship which will satisfy the required specification of the filter. The steps involved in any successful filter design may be summarised as follows:

(a) Specification of a suitable filter characteristic – for example, attenuation, phase shift, size, weight etc.
(b) Selection of an appropriate rational function necessary to realise the specification.
(c) Analysis of the selected circuit (equivalent to the required transfer function) and the calculation of the component values of the chosen filter circuit.
(d) Preliminary computer analysis of the design using SPICE or any CAD (Computer Aided Design) package, followed by
(e) the construction and testing of the chosen circuit.

The second step usually involves what is often described as 'the approximation technique'. The filter specification, in the form of pass-band gain, transition bandwidth, stop-band attenuation and cut-off frequency, is normally given in the frequency domain. The selection of a realisable mathematical relationship which approximates to the given specification usually involves considerable algebraic manipulation. Two of the most popular relationships are considered here in a simplified form. The student is referred to the Bibliography for books covering the theory in greater detail.

From the outset it should be noted that almost all filter circuits are derived from the equivalent low-pass prototype. A typical *ideal* low-pass characteristic is shown in figure 2.1.

The requirements of the specification are that all frequencies below ω_c (cut-off) are passed with no attenuation and above ω_c are stopped with infinite attenuation. An additional requirement is that the phase $\phi(j\omega)$ should be a linear function of ω. An 'actual' magnitude characteristic is shown in figure 2.2.

It should be stressed at this point that the technique applied to filter design is one of choosing a filter response characteristic which closely approximates to the ideal and which finally results in a circuit realisation that contains a manageable number of components.

After the circuit has been realised using nominal component values to ascertain the correctness of the design, it is desirable to introduce tolerances within the components to assess whether or not the original speciication will still be met in a practical design.

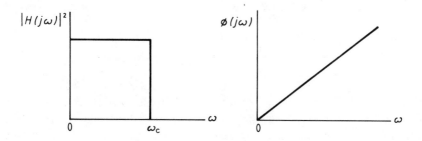

Figure 2.1   Ideal gain/phase responses

Generally, however, these parasitics, as they are often called, may be ignored in favour of completing an uncomplicated design, although sometimes they may be included within the design.

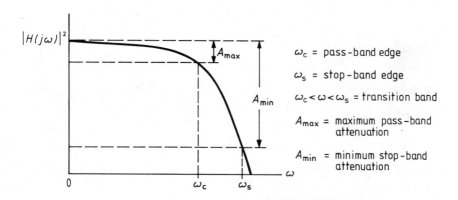

Figure 2.2   An actual gain response

2.2 The Butterworth approximation

The Butterworth characteristic takes the form described by the equation

$$|H(j\omega)|^2 = \frac{1}{1 + F(\omega^2)} \tag{2.1}$$

where $F(\omega^2) \gg 1$ for $\omega \gg \omega_c$ and $0 \leqslant F(\omega^2) \leqslant 1$ for $0 \leqslant \omega < 1$, so that in the pass-band $|H(j\omega)| \approx 1$ and in the stop-band $|H(j\omega)| \approx 0$.

Obviously the quality of the approximation depends on $F(\omega^2)$ which must be defined so as to make the approximation as accurate as possible. Butterworth proposed that

$$F(\omega^2) = \epsilon^2 \omega^{2n}; \ n = 1, 2, 3 \ldots$$

where, in the Butterworth response, $\epsilon = 1$ and is called the *ripple factor*. Being unity, there are no ripples in the pass-band of the Butterworth response. If the function $F(\omega^2)$ is expanded using a Taylor series and derivates found for this expansion, then for the condition around $\omega = 0$ all the derivatives, except one, equal zero. This is often referred to as the *maximally flat* condition.

Insertion into equation (2.1) yields

$$|H(j\omega)| = \frac{1}{(1 + \epsilon^2 \omega^{2n})^{\frac{1}{2}}} \tag{2.2}$$

Equation (2.2) defines the AMPLITUDE response of an nth-order Butterworth filter and this approximation produces a particularly flat response close to $\omega = 0$. At $\omega = \omega_c = 1$ (normalised):

$$|H(j\omega)| = \frac{1}{(1 + \epsilon^2)^{\frac{1}{2}}} \tag{2.3}$$

The approximation approaches the ideal response as n increases, as shown in figure 2.3.

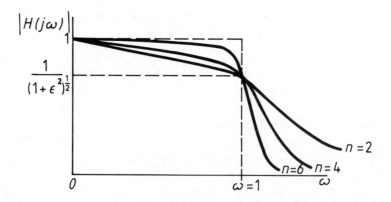

Figure 2.3 Butterworth gain response

The so-called *maximally flat* response is typical of the Butterworth approximation. The ripple factor ϵ has no effect on the pass-band response since in all of the cases the pass-band attenuation is taken to be 3 dB; this yields $\epsilon = 1$ from equation (2.6). Equation (2.2) may be written

$$|H(j\omega)| = \frac{1}{(1 + \omega^{2n})^{\frac{1}{2}}}$$

and for $\omega \gg 1$ this becomes $|H(j\omega)| \approx 1/\omega^n$ and the stop-band attenuation is then

$$\alpha_s = 20\log_{10}|H(j\omega)| = -20n\log_{10}\omega \qquad (2.4)$$

Equation (2.4) shows that the attenuation 'rolls-off' at $20n$ dB/decade change in frequency (giving the slope of the response characteristic in the transition band).

The pass-band attenuation is obtained from

$$\alpha_p = 20\log_{10}|H(j\omega)| = 20\log_{10}\frac{1}{(1 + \omega^{2n})^{\frac{1}{2}}}$$

$$= -10\log_{10}(1 + \omega^{2n}) \qquad (2.5)$$

It was stated that $\epsilon = 1$, but equation (2.5) can be used to obtain the ripple factor for given filter specifications; an example would be for the Chebyshev response, to be discussed later.

From equation (2.5) the pass-band attenuation may be obtained in the form

$$\alpha_p = -10\log_{10}(1 + \epsilon^2\omega^{2n})$$ and at $\omega = 1$ this should be A_{MAX}, that is

$$|A_{MAX}| = 10\log_{10}(1 + \epsilon^2)$$ yielding

$$\epsilon = (10^{A\,MAX/10} - 1)^{\frac{1}{2}} \qquad (2.6)$$

Furthermore, it is important that the order (n) of the filter be obtained from the given specifications, and

$$|A_{MIN}| = 10\log_{10}(1 + \epsilon^2\omega^{2n}) \qquad (2.7)$$

Substituting for ϵ gives

$$A_{MIN} = 10\log_{10}\left[1 + \{10^{A\,MAX/10} - 1\}(\omega^{2n})\right]$$

and

$$\omega^{2n} = \left\{\frac{10^{A\,MIN/10} - 1}{10^{A\,MAX/10} - 1}\right\} \quad \text{and} \quad \omega = (\omega_s/\omega_c)$$

Taking logarithms, we obtain

$$n = \frac{1}{2\log_{10}\left(\dfrac{\omega_s}{\omega_c}\right)} \times \log_{10}\left\{\frac{10^{A\,MIN/10} - 1}{10^{A\,MAX/10} - 1}\right\} \qquad (2.8)$$

Using equations (2.6) and (2.8), the ripple factor (width) and the order n may be calculated. For example

for $A_{MAX} = 3$ dB in the Butterworth case

$$\epsilon = (e^{0.3} - 1)^{\frac{1}{2}} = 1$$

Since $F(\omega^2)$ is a polynomial in ω, it follows that the roots of the denominator of equation (2.2) must be obtained since they dictate the nature of the response for any order n.

Using $|H(j\omega)| = \dfrac{1}{(1 + \omega^{2n})^{\frac{1}{2}}}$ with $\epsilon = 1$

and writing $s = j\omega$, then

$$|H(s)|^2 = \frac{1}{1 - (s^2)^n}$$

Solving for the roots of $1 - (s^2)^n = 0$ yields:

for n even

$s^{2n} = -1 = \exp[j(2k-1)\pi]$ whose $2n$ roots are:

$$p_k = \exp\left\{ \frac{j(2k-1)\pi}{2n} \right\} \qquad k = 1, 2 \ldots 2n$$

or $\quad p_k = \cos\left[\dfrac{(2k-1)\pi}{2n} + j\sin\dfrac{(2k-1)\pi}{2n} \right]$

for n odd

$s^{2n} = 1 = \exp(j2k\pi)$ whose $2n$ roots are:

$$p_k = \exp\left[j\frac{k}{n}\pi \right] \qquad k = 0, 1 \ldots (2n-1)$$

or $\quad p_k = \cos\dfrac{k}{n}\pi + j\sin\dfrac{k}{n}\pi$

The transfer functions may be summarised:

for n even

$$|H(s)|^2 = \frac{1}{\prod\limits_{k=1}^{n/2} (s^2 + 2\cos\theta_k s + 1)}$$

$$\theta_k = \frac{(2k-1)\pi}{2n}$$

for n odd

$$|H(s)|^2 = \frac{1}{(s+1)^{(n-1)/2} \prod\limits_{k=1}^{} (s^2 + 2\cos\theta_k s + 1)}$$

$$\theta_k = \frac{k\pi}{n}$$

For example, $n = 2$:

$$|H(s)|^2 = \frac{1}{s^2 + 2\cos\dfrac{\pi}{4}s + 1}$$

$$= \frac{1}{s^2 + \sqrt{2}s + 1}$$

For example, $n = 3$:

$$|H(s)|^2 = \frac{1}{(s + 1)\left(s^2 + 2\cos\dfrac{\pi s}{3} + 1\right)}$$

$$= \frac{1}{(s^3 + 2s^2 + 2s + 1)}$$

A table of factors may now be constructed and is shown in table 2.1 up to and including the eighth-order polynomial.

Table 2.1 Table of factors for Butterworth polynomial

n	Roots	Factors	Polynomial
1	-1	$s + 1$	$s + 1$
2	$-\cos\dfrac{\pi}{4} \pm j\sin\dfrac{\pi}{4}$	$s^2 + \sqrt{2}s + 1$	$s^2 + \sqrt{2}s + 1$
3	$-1, -\cos\dfrac{\pi}{3} \pm j\sin\dfrac{\pi}{3}$	$(s + 1), (s^2 + s + 1)$	$s^3 + 2s^2 + 2s + 1$
4	$-\cos k\dfrac{\pi}{8} \pm j\sin k\dfrac{\pi}{8}$ $k = 1, 3 \ldots$	$(s^2 + 0.765s + 1), (s^2 + 1.848s + 1)$	$s^4 + 2.613s^3 + 3.414s^2$ $+ 2.613s + 1$
5	$-1, -\cos k\dfrac{\pi}{5} \pm j\sin k\dfrac{\pi}{5}$ $k = 1, 2$	$(s + 1), (s^2 + 0.618s + 1),$ $(s^2 + 1.618s + 1)$	$s^5 + 3.236s^4 + 5.236s^3$ $+ 5.236s^2 + 3.236s + 1$
6	$-\cos k\dfrac{\pi}{12} \pm j\sin k\dfrac{\pi}{12}$ $k = 1, 3, 5$	$(s^2 + 0.518s + 1), (s^2 + \sqrt{2}s + 1),$ $(s^2 + 1.932s + 1)$	$s^6 + 3.864s^5 + 7.464s^4$ $+ 9.142s^3 + 7.464s^2$ $+ 3.864s + 1$
7	$-1, -\cos k\dfrac{\pi}{7} \pm j\sin k\dfrac{\pi}{7}$ $k = 1, 2, 3$	$(s + 1), (s^2 + 0.44s + 1),$ $(s^2 + 1.247s^2 + 1), (s^2 + 1.802s + 1)$	$s^7 + 4.494s^6 + 10.098s^5$ $+ 14.592s^4 + 14.592s^3$ $+ 10.098s^2 + 4.494s + 1$
8	$-\cos k\dfrac{\pi}{16} \pm j\sin k\dfrac{\pi}{16}$ $k = 1, 3, 5, 7$	$(s^2 + 0.399s + 1), (s^2 + 1.111s + 1),$ $(s^2 + 1.166s + 1), (s^2 + 1.962s + 1$	$s^8 + 5.126s^7 + 13.137s^6$ $+ 21.846s^5 + 25.688s^4$ $+ 21.846s^3 + 13.137s^2$ $+ 5.126s + 1$

An example will now be considered with which to illustrate how such a table of factors, as shown in table 2.1, may be derived.

Example 2.1

A Butterworth low-pass filter is required to provide the following specifications: pass band up to 10 kHz with 3 dB of loss, stop band to be at least 100 dB down at 100 kHz. Calculate the order of the filter and the roots of the Butterworth polynomial.

Here $A_{MAX} = 3$ dB, $A_{MIN} = 100$ dB, $\epsilon = 1$.

$$n = \frac{1}{2\log_{10} 10} \log_{10} \left\{ \frac{10^{10} - 1}{10^3 - 1} \right\}$$

$$= \tfrac{1}{2} \times \log_{10} \frac{10^{10}}{0.995} = 5$$

We have an odd function, hence $\theta_n = k\pi/n$ and a root on the negative real axis.

$$\frac{k\pi}{n} = \theta_n$$

$k = 0 \quad \theta_0 = 0$

$k = 1 \quad \theta_1 = \pi/5 = 36°$

$k = 2 \quad \theta_2 = 2\pi/5 = 72°$

Roots are $(s + \sigma_0)(s + \sigma_1 + j\omega_1)(s + \sigma_1 - j\omega_1)(s + \sigma_2 + j\omega_2)(s + \sigma_2 - j\omega_2)$.

$$\sigma_1 \pm j\omega_1 = \cos 36° \pm j\sin 36° = 0.809 \pm j0.587$$

$$\sigma_2 \pm j\omega_2 = \cos 72° \pm j\sin 72° = 0.309 \pm j0.951$$

$$\sigma_0 = \cos 0° \pm j\sin 0° = 1$$

And

$$|H(s)|^2 = \frac{1}{(s + 1)(s^2 + 1.618s + 1)(s^2 + 0.618s + 1)}$$

Example 2.2

It is required to determine the order of a Butterworth filter having the following specification: $A_{MIN} = 20$ dB, $A_{MAX} = 3$ dB, $\omega_s/\omega_c = 1.5$.

From equation (2.8):

$$n = \frac{1}{2\log_{10} 1.5} \times \log_{10} \left\{ \frac{10^2 - 1}{10^{0.3} - 1} \right\} = 5.67$$

Choose $n = 6$ and from equations (2.6) and (2.7) $\epsilon = (10^{0.3} - 1)^{\frac{1}{2}} = 1$ which is as expected for Butterworth.

$$A_{MIN} = 10\log_{10}(1 + 1^2(1.5)^{12}) = 10 \times 2.12 = 21.2 \text{ dB}$$

which meets the specification

2.3 The Chebyshev approximation

It was seen that the Butterworth form gives a good approximation to the ideal around ($\omega = 0$) but is less than ideal in the vicinity of the cut-off point ($\omega = \omega_c$).

This is because $F(\omega^2) = \omega^{2n}$ is chosen such that the zeros of $F(\omega^2)$ are at one point ($\omega = 0$). An alternative approximation, called the Chebyshev approximation, spreads the zeros of $F(\omega^2)$ across the pass band and constrains $H(j\omega)$ to attain its maximum value at a number of points within the pass band. It will be seen that the ripple factor ϵ affects the way the response behaves in the pass band. Depending on the order of the filter, pass-band ripples occur having equal heights and variable frequency, the ripple height (or width) being determined by the ripple factor.

Here the Chebyshev polynomial is defined by

$$F(\omega^2) = \epsilon^2 C_n^2(\omega) \text{ and} \tag{2.9}$$

$$|H(j\omega)| = \frac{1}{(1 + \epsilon^2 C_n^2(\omega))^{\frac{1}{2}}} \text{ where} \tag{2.10}$$

$$\epsilon = (10^{A\text{MAX}/10} - 1)^{\frac{1}{2}} \text{ as before}$$

The polynomial $C_n(\omega)$ is defined as

$$\begin{aligned} C_n(\omega) &= \cos(n\cos^{-1}\omega) \quad \text{for } \omega \leqslant 1 \\ C_n(\omega) &= \cosh(n\cosh^{-1}\omega) \quad \text{for } \omega > 1 \end{aligned} \tag{2.11}$$

$C_n(\omega)$ exhibits an equal ripple property in the pass band.

$$n = 0; \ C_0(\omega) = 1$$

$$n = 1; \ C_1(\omega) = \omega$$

To obtain the higher-order polynomials we use the relationship

$$C_{n+1}(\omega) = 2\omega C_n(\omega) - C_{n-1}(\omega) \tag{2.12}$$

Table 2.2 may be constructed using equations (2.11) and (2.12).

Table 2.2 A table of Chebyshev polynomials

n	$C_n(\omega)$
0	1
1	ω
2	$2\omega^2 - 1$
3	$4\omega^3 - 3\omega$
4	$8\omega^4 - 8\omega^2 + 1$
5	$16\omega^5 - 20\omega^3 + 5\omega$
6	$32\omega^6 - 48\omega^4 + 18\omega^2 - 1$
7	$64\omega^7 - 112\omega^5 + 56\omega^3 - 7\omega$
8	$128\omega^8 - 256\omega^6 + 160\omega^4 - 32\omega^2 + 1$

If these functions are plotted we see the equal ripple characteristics for $\omega \leqslant 1$, and two examples are given in figure 2.4.

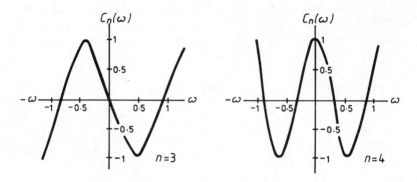

Figure 2.4   Plots of $C_n(\omega)$ for $n = 3$, $n = 4$

Consideration of equation (2.10), repeated here

$$|H(j\omega)| = \frac{1}{(1 + \epsilon^2 C_n^2(\omega))^{\frac{1}{2}}}$$

shows that within the pass band ($0 < \omega \leqslant 1$), $C_n^2(\omega) = 1$ and the function will vary between $1/(1 + \epsilon^2)^{\frac{1}{2}}$ and 1.

Within the transition band and through the stop band, however, $C_n^2(\omega)$ increases with ω such that for $\epsilon^2 C_n^2(\omega) \gg 1$, $|H(j\omega)|$ decreases rapidly. These features are shown in figure 2.5 from which the *ripple width* may be determined:

$$\text{ripple width} = 1 - \frac{1}{(1 + \epsilon^2)^{\frac{1}{2}}} \tag{2.13}$$

To determine the order n of the filter from given specifications, we note from equation (2.10)

$$A_{\text{MIN}} = 10\log_{10}[1 + \epsilon^2 C_n^2(\omega)]$$

for the edge of the stop band at $\omega = \omega_s/\omega_c$ normalised. From which

$$\epsilon^2 C_n^2(\omega) = \epsilon^2(\cosh(n\cosh^{-1}\omega))^2 = 10^{A\,\text{MIN}/10} - 1 \quad \text{giving}$$

$$|A_{\text{MIN}}| = 10\log_{10}[1 + \epsilon^2\{\cosh(n\cosh^{-1}\omega)\}^2] \tag{2.14}$$

and $\epsilon^2 = 10^{A\,\text{MAX}/10} - 1$

Which upon insertion into (2.14) gives

$$\cosh(n\cosh^{-1}\omega) = \left[\frac{10^{A\,\text{MIN}/10} - 1}{10^{A\,\text{MAX}/10} - 1}\right]^{\frac{1}{2}}$$

and which finally yields

$$n = \frac{\cosh^{-1}[(10^{A\,\text{MIN}/10} - 1)/(10^{A\,\text{MAX}/10} - 1)]^{\frac{1}{2}}}{\cosh^{-1}\left(\dfrac{\omega_s}{\omega_c}\right)} \qquad (2.15)$$

Also since for the pass band pass $C_n^2(\omega) = 1$:

$$|A_{\text{MAX}}| = 10\log_{10}[1 + \epsilon^2] \qquad (2.16)$$

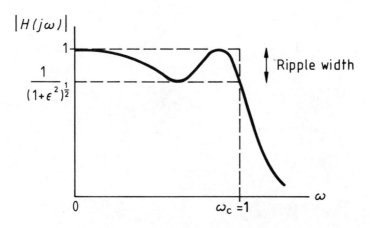

Figure 2.5   Effect of ripple width

Unlike the Butterworth approximation, it cannot be assumed that the 3 dB or *half-power* frequency is the cut-off frequency. The end of the pass band or ripple band is always at $\omega = \omega_c$. For the condition $\epsilon = 1$ gives a ripple width of 3 dB and $\omega = \omega_c$ as for the Butterworth approximation.

To determine the half-power frequency we note that $\epsilon^2 C_n^2(\omega) = 1$ which for $\omega = \omega_{3\text{dB}}$ gives

$$C_n(\omega_{3\text{dB}}) = \frac{1}{\epsilon} = \cosh(n\cosh^{-1}\omega_{3\text{dB}})$$

which yields $\cosh^{-1}\omega_{3\text{dB}} = \dfrac{1}{n}\cosh^{-1}\left(\dfrac{1}{\epsilon}\right)$ and finally

$$\omega_{3\text{dB}} = \cosh\left[\frac{1}{n}\cosh^{-1}\left(\frac{1}{\epsilon}\right)\right] \qquad (2.17)$$

Finally it is necessary to determine the roots of the Chebyshev polynomial for various values of the order n of the filter.

The roots are obtained from the expression

$$1 + \epsilon^2 C_n^2(\omega) = 0 \tag{2.18}$$

and are $s_k = \sigma_k \pm j\omega_k, \quad k = 0, 1, 2 \ldots 2n - 1$

where $\sigma_k = \sinh a \sin(2k + 1) \dfrac{\pi}{2n}$

$$\omega_k = \cosh a \cos(2k + 1) \dfrac{\pi}{2n} \tag{2.19}$$

$$\theta_k = (2k + 1) \dfrac{\pi}{2n}$$

for $k = 0, 1, 2 \ldots 2n - 1$ and

$$a = \frac{1}{n} \sinh^{-1}\left(\frac{1}{\epsilon}\right) \tag{2.20}$$

Using (2.19) and writing

$$\frac{\sigma_k}{\sinh a} = \sin(2k + 1) \frac{\pi}{2n}$$

$$\frac{\omega_k}{\cosh a} = \cos(2k + 1) \frac{\pi}{2n}$$

then

$$\left(\frac{\sigma_k}{\sinh a}\right)^2 + \left(\frac{\omega_k}{\cosh a}\right)^2 = 1$$

which indicates that the poles (roots) of $|H(s)|^2$ lie on an ellipse. It is interesting to note that those of the Butterworth polynomial lie on a circle. Furthermore, it should be noted that the Chebyshev poles depend on the ripple factor ϵ, and tables are constructed for various values of ripple width. These are shown in table 2.3.

It may further be shown for the special case when $\omega = \omega_c$, that is for $\epsilon = 1$, the Chebyshev response has a slope which is n times that of the Butterworth response.

If, briefly, we consider the relative phase responses shown in figure 2.6, we may conclude that the choice of response type will be governed by: high attenuation in the stop band and sharper roll-off in the vicinity of the cut-off frequency as opposed to a non-linear phase characteristic in the stop band.

Table 2.3 Poles of the Chebyshev polynomial for values of A_{MAX}

A_{MAX} = 0.1 dB	
$n = 1$	$s + 6.552$
$n = 2$	$s^2 + 2.3724s + 3.314$
$n = 3$	$(s^2 + 0.9694s + 1.6897)(s + 0.9694)$
$n = 4$	$(s^2 + 0.5283s + 1.33)(s^2 + 1.2754s + 0.6229)$
A_{MAX} = 0.5 dB	
$n = 1$	$s + 2.8627$
$n = 2$	$s^2 + 1.4256s + 1.5162$
$n = 3$	$(s^2 + 0.6264s + 1.1424)(s + 0.6264)$
$n = 4$	$(s^2 + 0.3507s + 1.0635)(s^2 + 0.8466s + 0.3564)$
A_{MAX} = 1 dB	
$n = 1$	$s + 1.9652$
$n = 2$	$s^2 + 1.0977s + 1.1025$
$n = 3$	$(s^2 + 0.4941s + 0.9942)(s + 0.4941)$
$n = 4$	$(s^2 + 0.279s + 0.9865)(s^2 + 0.6737s + 0.2794)$

Example 2.3

The specification for a Chebyshev filter is given by

$$A_{MIN} = 20 \text{ dB}, \quad A_{MAX} = 3 \text{ dB}, \quad \omega_s/\omega_c = 1.5$$

Obtain the order of the filter and confirm that the stop band requirement is fulfilled.

From equation (2.15):

$$n = \frac{\cosh^{-1}[(10^2 - 1)/(10^{0.3} - 1)]^{\frac{1}{2}}}{\cosh^{-1} 1.5} = 3.11$$

Choose $n = 4$ and from equations (2.6) and (2.14)

$$\epsilon = (10^{0.3} - 1)^{\frac{1}{2}} = 1$$

giving

$$A_{MIN} = 10\log_{10}[1 + (1)^2 \{\cosh(4\cosh^{-1}1.5)\}^2]$$

$$= 10\log_{10}[1 + \cosh(3.85)]$$

$$= \underline{27.4 \text{ dB}}$$

Note that this fourth-order Chebyshev circuit satisfies the requirements better than the previously considered sixth-order Butterworth circuit, at the expense of some pass-band ripple.

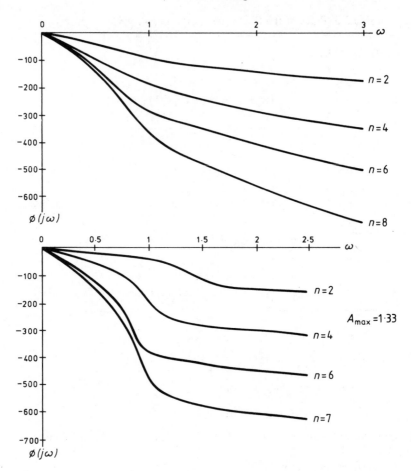

Figure 2.6　Butterworth and Chebyshev phase responses

Example 2.4

A Chebyshev filter is required to provide the following specification:

　　pass-band maximum ripple width 0.5 dB up to 3 kHz

　　stop-band minimum to be 60 dB at 30 kHz

Calculate the order of the filter and the roots of the polynomial.
　Given A_{MAX} = 0.5 dB, $A_{MIN} > 60$ dB.

$$n = \frac{\cosh^{-1}[(10^6 - 1)/(10^{0.05} - 1)]^{\frac{1}{2}}}{\cosh^{-1} 30/3} = 2.88$$

Take $n = 3$: using equation (2.6) we find that $\epsilon = 0.35$.

$$a = \frac{1}{3} \sinh^{-1} \left(\frac{1}{0.35} \right) = 0.59$$

$$\theta_k = (2k + 1) \frac{\pi}{2n}$$

giving

$$\theta_0 = \frac{\pi}{6} = 30°$$
$$\theta_1 = 90°$$

There are poles at $s_1 = -\sinh a = -\sinh 0.59 = -0.625$

$$s_2 = -0.625 \sin 30° \pm j1.18 \cos 30°$$
$$= -0.313 \pm j1.022$$
$$s_3 = -0.625 \sin 90° \pm j1.18 \cos 90°$$
$$= -0.625 \text{ (repeated root)}$$

The transfer function takes the form

$$| H(s)|^2 = \frac{1}{(s + 0.625)(s + 0.313 + j1.022)(s + 0.313 - j1.022)}$$

$$= \frac{1}{(s + 0.625)(s^2 + 0.626s + 1.142}$$

This result should be compared with the data given in table 2.3. It should again be noted that we have also obtained important design coefficients which will later be used in the design of filters having specified ripple widths.

We may write

$$| H(s)|^2 = \frac{1}{(s + b_0)(s^2 + b_1 s + b_2)}$$

where $b_0 = 0.625$, $b_1 = 0.625$, $b_2 = 1.142$.

Table 2.4 shows such coefficients for various order n of the filter and different pass-band ripple widths.

2.4 Conclusions

We may conclude that the Chebyshev response gives a superior pass-band response, particularly if the ripple width is not too large, than does the Butterworth response. Unfortunately this superiority in amplitude response must be traded

Table 2.4 Chebyshev coefficients for various orders and maximum ripple widths

Order n	A_{MAX} (dB)	b_0	b_1
2	0.5	1.516	1.426
	1.0	1.1025	1.0977
	3.0	0.708	0.645
3	0.5	1.142	0.626
		0.626	
	1.0	0.994	0.494
		0.494	
	3.0	0.839	0.298
		0.298	
4	0.5	1.0635	0.3507
		0.3564	0.8466
	1.0	0.9865	0.279
		0.279	0.6737
	3.0	0.903	0.1703
		0.1959	0.4112

against a poorer, less linear, phase response, up to the cut-off frequency. The Butterworth phase response is far more linear, as shown in figure 2.6. A further point is that in the vicinity of the cut-off frequency, the roll-off of the response for the Chebyshev is n times steeper than that for the Butterworth.

Problems

2.1. A Butterworth low-pass filter has the following specifications: pass band up to 100 Hz with 3 dB of loss, stop band to be at least 80 dB down at 1 kHz. Calculate the order of the filter, the roots of the Butterworth polynomial, and the values of the pass-band and stop-band loss.
[*Ans.* 4; 0.765, 1, 1.848, 1; 3 dB, 80 dB]

2.2. Obtain the normalised Chebyshev coefficients for a fourth-order, 1 dB ripple width, low-pass filter.
[*Ans.* $\epsilon = 0.509$; 0.279, 0.9865; 0.6737, 0.2793]

2.3. A Chebyshev low-pass filter is required to have a pass-band ripple width of 0.5 dB from dc to 150 Hz, and to be at least 100 dB down at 1.5 kHz. Calculate the order of the filter and the Chebyshev polynomial. Check also the values of the pass-band and stop-band loss.
[*Ans.* 5; 0.362; 0.586, 0.476; 0.224, 1.035; 0.5 dB, 114.8 dB]

3 The Operational Amplifier

3.1 Introduction

It is useful to consider at this stage the basic theory concerning integrated circuit operational amplifiers, since they are the foundation of most of the filter circuits to be studied. The reader should refer to the more specialised texts for a fuller treatment of the subject matter.

The operational amplifier (op-amp) represents a basic integrated circuit building block and is a broad-band, high-gain, differential input device. Initially, the op-amp was applied to the solution of such mathematical functions as integration, differentiation and summing by electronic means. Nowadays, the applications of the device are legion.

At the present time, the availability of inexpensive integrated circuit amplifiers, either in single, dual or quad form, has made the packaged op-amp a useful replacement for discrete amplifiers in many active circuit realisations.

The versatility of the op-amp is improved by using negative feedback which improves stability of gain, reduces output impedance and improves linearity of operation. Positive and negative power supplies are required, both referenced to earth potential.

A popular and inexpensive op-amp is the μA741 which features offset-null capability, protection against short-circuit, and built-in frequency compensation, which results in unity gain bandwidth of 1 MHz. The μA748 is similar to the μA741 but features externally selected frequency compensation through the choice of a single capacitor.

Dual and quad op-amps have been introduced to reduce cost and space; in particular, the quad op-amp has found particular favour with filter designers. A typical specification and op-amp pin layout is shown in figure 3.1 for the popular μA741 monolithic amplifier.

3.2 The ideal op-amp

A simplified model for representing an op-amp is shown in figure 3.2 where R_i, R_o are the input and output impedances and A is the open-loop gain of the op-amp.

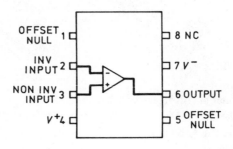

ELECTRICAL CHARACTERISTICS (V_S = ±15 V, T_A = 25°C unless otherwise specified)

Parameters		Conditions	Min.	Typ.	Max.	Units
Input offset voltage		$R_S < 10\,k\Omega$		2.0	6.0	mV
Input offset current				20	200	nA
Input bias current				80	500	nA
Input resistance			0.3	2.0		MΩ
Input capacitance				1.4		pF
Offset voltage adjustment range				±15		mV
Input voltage range			±12	±13		V
Common mode rejection ratio		$R_S < 10\,k\Omega$	70	90		dB
Supply voltage rejection ratio		$R_S < 10\,k\Omega$		30	150	μV/V
Large signal voltage rain		$R_L > 2\,k\Omega$, $V_{OUT} = \pm10$ V	20 000	200 000		
Output voltage swing		$R_L > 10\,k\Omega$	±12	±14		V
		$R_L > 2\,k\Omega$	±10	±13		V
Output resistance				75		Ω
Output short-circuit current				25		mA
Supply current				1.7	2.8	mA
Power consumption				50	85	mW
Transient response (unity gain)	Risetime	V_{IN} = 20 mV, R_L = 2 kΩ, $C_L < 100$ pF		0.3		μs
	Overshoot			5.0		%
Slew rate		$R_L > 2\,k\Omega$		0.5		V/μs

The following specifications apply for 0°C $< T_A <$ ±70°C:

Input offset voltage					7.5	mV
Input offset current					300	nA
Input bias current					800	nA
Large signal voltage gain		$R_L > 2\,k\Omega$, $V_{OUT} = \pm10$ V	15 000			
Output voltage swing		$R_L > 2\,k\Omega$	±10	±13		V

Figure 3.1 μA741 specification

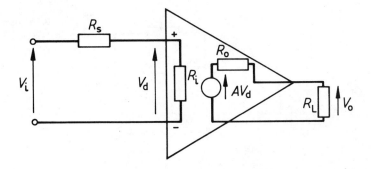

Figure 3.2   Basic voltage amplification circuit

Analysis of the circuit yields several equations:

$$V_d = \frac{R_i}{R_i + R_s} \, V_i \tag{3.1}$$

$$V_o = \frac{R_L}{R_o + R_L} \, A V_d \tag{3.2}$$

or

$$\frac{V_o}{V_i} = \frac{A R_i R_L}{(R_i + R_s)(R_o + R_L)} = \frac{A}{\left(1 + \dfrac{R_s}{R_i}\right)\left(1 + \dfrac{R_o}{R_L}\right)}$$

Normally $R_i \gg R_s$, $R_L \gg R_o$ and we obtain

$$\frac{V_o}{V_i} \approx A \tag{3.3}$$

This latter result shows that under conditions approaching ideal, the circuit behaves as an ideal *voltage-controlled-voltage-source* (VCVS). We conclude, therefore, that an ideal op-amp has the following: infinite input impedance, zero output impedance, infinite open-loop gain and infinite bandwidth.

3.3 Two important op-amp configurations

We now consider two basic circuits which will be used in subsequent chapters.

3.3.1 The inverting circuit

The basic connection is shown in figure 3.3 where the input is connected via resistor R_1 to the inverting terminal, the non-inverting terminal being connected to earth. Feedback is applied via resistor R_2.

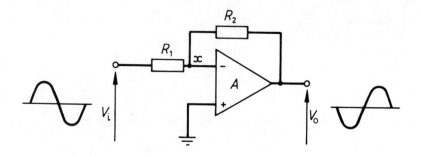

Figure 3.3 Inverting amplifier

Nodal analysis at point x yields

$$\frac{V_x - V_i}{R_1} + \frac{V_x - V_o}{R_2} = 0 \quad \text{also} \quad V_o = -A V_x$$

Eliminating V_x we obtain after a little manipulation, the expression

$$\frac{V_o}{V_i} = - \frac{A R_2}{(1 + A)\left(R_1 + \dfrac{R_2}{1 + A}\right)} \tag{3.4}$$

Normally the open-loop gain A is large so that the closed-loop gain becomes, to a good approximation

$$\frac{V_o}{V_i} = - \frac{R_2}{R_1} \tag{3.5}$$

The negative sign implies inversion of the input signal (a $180°$ phase shift between input and output).

3.3.2 The non-inverting circuit

In this configuration, the signal is applied to the non-inverting terminal and the inverting terminal is connected to earth via resistor R_1. Feedback is again supplied through resistor R_2 as shown in figure 3.4.

Applying nodal analysis at point x yields

$$\frac{V_x}{R_1} + \frac{V_x - V_o}{R_2} = 0 \quad \text{also} \quad A(V_i - V_x) = V_o \text{ or}$$

$$V_x = V_i - \frac{V_o}{A}$$

Eliminating V_x we obtain after manipulation

$$\frac{V_o}{V_i} = \frac{A}{(1+A)} \times \frac{R_1 + R_2}{R_1 + \dfrac{R_2}{1+A}} \tag{3.6}$$

If we assume a large open-loop gain A, then the expression for the closed-loop gain reduces to

$$\frac{V_o}{V_i} = 1 + \frac{R_2}{R_1} \tag{3.7}$$

Equation (3.7) shows that there is no phase difference between input and output. We find that the inverting circuit has a gain with feedback, fixed by the ratio of two resistors, so that in theory, closed-loop gains less than unity may be achieved, although we shall see that this condition has little practical significance for filter circuits.

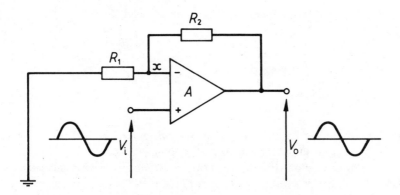

Figure 3.4 Non-inverting amplifier

The open-loop gain may be assumed to be high and the output impedance low. The input impedance, however, is fixed by the value of the resistor R_1 and consequently care must be exercised for some applications.

The non-inverting circuit has a minimum gain of unity when feedback is applied. The open-loop gain is also high and the circuit exhibits a low output impedance. Unlike the inverting circuit, the non-inverting configuration produces a high input impedance, AR_i (notice that now node x is not at virtual earth). An interesting modification to the circuit may be made by making R_2 a short-circuit and R_1 an open-circuit; the gain becomes unity $[V_o/V_i = 1 + (0/\infty)]$ and the circuit is then described as a *unity gain buffer* or *voltage follower* circuit.

3.4 The non-ideal op-amp

It is possible to construct op-amps to give a good approximation to the ideal within prescribed frequency limits. However, the gain of the op-amp is a function of the frequency of the applied signal and the actual behaviour can only approach the ideal over a specified range of frequency. Consider the simple op-amp model shown in figure 3.5.

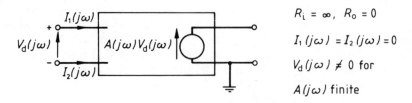

$$R_i = \infty, \quad R_o = 0$$

$$I_1(j\omega) = I_2(j\omega) = 0$$

$$V_d(j\omega) \neq 0 \text{ for}$$

$$A(j\omega) \text{ finite}$$

Figure 3.5 Frequency dependent model of op-amp

If we assume $A(j\omega)$ takes the form

$$A(j\omega) = \frac{A_0\omega_a}{j\omega + \omega_a} \qquad (3.8)$$

equation (3.8) describes the so-called *one-pole roll-off* model of the op-amp which has the gain and phase characteristics shown in figure 3.6.

For a given op-amp, the *gain–bandwidth product* is a constant value. This means that if a large open-loop gain is required then the working bandwidth will be narrow. Conversely, if the open-loop gain requirement is to be low then a much wider bandwidth will be obtained.

For example:

$$A_0 = 10^5, \omega_a = 2\pi \times 10$$

$$GB = 2\pi \times 10^6$$

For $A_0 = 10^3$, bandwidth $= \dfrac{2\pi \times 10^6}{10^3} = 2\pi \times 10^3$ rad/s

For $A_0 = 10$, bandwidth $= \dfrac{2\pi \times 10^6}{10} = 2\pi \times 10^5$ rad/s

The reader is directed to problem **3.4** where a different view of the bandwidth of an inverting circuit is proposed.

We may conclude that the gain–bandwidth product is an important item of data when selecting an op-amp for a particular design:

$$GB = A_0 \omega_a \qquad (3.9)$$

As we have seen, typical values would be $A_0 = 10^5$, $\omega_a = 2\pi \times 10$ rad/s, which when inserted into the magnitude expression obtained from equation (3.8) yields

$$|A(jA_0\omega_a)| = \frac{A_0\omega_a}{((A_0\omega_a)^2 + \omega_a^2)^{0.5}} \approx 1, \ \omega = A_0\omega_a$$

At very low frequencies ($\omega \to 0$) the gain becomes

$$|A(j0)| = \frac{A_0\omega_a}{\omega_a} = A_0$$

and for the values of A_0, ω_a quoted, it is seen that the gain is

$$|A(j0)| = 20\log_{10}10^5 = 100 \text{ dB}$$

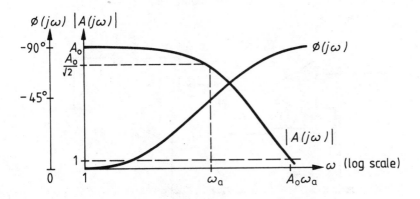

Figure 3.6   Response of non-ideal op-amp

We see that the gain is 100 dB down when the frequency is equal to the gain-bandwidth product. Note also that there is an additional 90° phase shift through the op-amp over the working frequency range.

It should finally be mentioned here that the quality of the output signal depends on the op-amp's ability to cope with the rate at which the output signal varies with time. This capability is referred to as the *slew rate* of the op-amp and is the maximum rate-of-change of the output voltage which the device can handle.

$$\text{slew rate} = \left.\frac{dV_o}{dt}\right|_{\max} \qquad (3.10)$$

Selection of op-amps with superior GB and SR is left to the designer, who must choose between cost and suitability of a device for his particular requirement. It should be noted that non-ideal op-amps have finite, (usually) high values of input resistance and non-zero values of output resistance. Other quantities of interest are: common-mode rejection, input offset voltages and currents etc.; this latter condition is of importance to the filter designer and will be briefly investigated here.

In the case of an ideal op-amp, the output voltage will be zero if both input voltages are zero. For a real op-amp however, there may be an output voltage even if both input voltages are zero. This output *offset* voltage is obtained by grounding each input terminal and measuring the output voltage. As an example, consider $A_o = 10^4$ and that a V_o of 5 V is measured; then

$$V_{\text{offset}} = \frac{5}{10^4} = 0.5 \text{ mV}$$

In order to illustrate the important analytical method of nodal analysis, consider the instrumentation amplifier of figure 3.7.

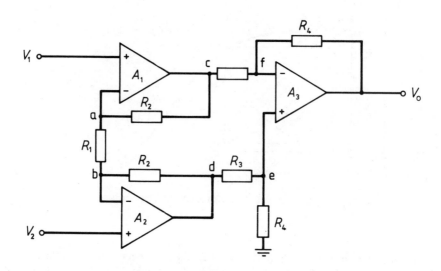

Figure 3.7   Instrumentation amplifier circuit

Example 3.1

The circuit shown in figure 3.7 is that of an instrumentation amplifier. Obtain the relationship between the output and the two inputs. Assume that the op-amps used have high open-loop gains and input impedance values.

Using nodal analysis at the labelled nodes, we obtain the following equations:

node a

$$\frac{V_a - V_c}{R_2} + \frac{V_a - V_b}{R_1} = 0, V_a\left(\frac{1}{R_1} + \frac{1}{R_2}\right) = \frac{V_c}{R_2} + \frac{V_b}{R_1} \tag{3.11}$$

node b

$$\frac{V_b - V_d}{R_2} + \frac{V_b - V_a}{R_1} = 0, V_b\left(\frac{1}{R_1} + \frac{1}{R_2}\right) = \frac{V_a}{R_1} + \frac{V_d}{R_2} \tag{3.12}$$

node c

$$\frac{V_c - V_f}{R_3} + \frac{V_c - V_a}{R_2} = 0, V_c\left(\frac{1}{R_2} + \frac{1}{R_3}\right) = \frac{V_a}{R_2} + \frac{V_f}{R_3} \tag{3.13}$$

node e

$$\frac{V_e}{R_4} + \frac{V_e - V_d}{R_3} = 0, V_e\left(\frac{1}{R_3} + \frac{1}{R_4}\right) = \frac{V_d}{R_3} \tag{3.14}$$

node f

$$\frac{V_f - V_o}{R_4} + \frac{V_f - V_c}{R_3} = 0, V_f\left(\frac{1}{R_3} + \frac{1}{R_4}\right) = \frac{V_o}{R_4} + \frac{V_c}{R_3} \tag{3.15}$$

Since a major assumption is that the gains of the op-amps are very high, then we may quote the following voltages: $V_b = V_2, V_a = V_1, V_e = V_f$.

From (3.15): $V_f = \dfrac{R_3}{R_3 + R_4} V_o + \dfrac{R_4}{R_3 + R_4} V_c = V_e$

From (3.14): $V_e = \dfrac{R_4}{R_3 + R_4} V_d$

Equating these two expressions yields:

$$V_d = \frac{R_3}{R_4} V_o + V_c \tag{3.16}$$

From (3.12): $V_c = \dfrac{R_1 + R_2}{R_1} V_2 - \dfrac{R_2}{R_1} V_1 - \dfrac{R_3}{R_4} V_o$ using (3.16)

From (3.11): $V_c = \dfrac{R_1 + R_2}{R_1} V_1 - \dfrac{R_2}{R_1} V_2$

Equating these two expressions finally yields:

$$V_o = \left(1 + \frac{2R_2}{R_1}\right)\frac{R_4}{R_3} (V_2 - V_1) \tag{3.17}$$

The resistor R_1 is usually made variable in order to adjust the gain. A typical instrumentation op-amp would have $A_o = 2 \times 10^5, R_i = 10^{12}$ ohms.

This circuit is an example of a differential-in and differential-out (A_1, A_2) amplifier with negative feedback and equalised amplification.

Problems

3.1. For the circuit shown in figure 3.8 it is required that $V_o = \dfrac{V_2}{3} - 2V_1$.

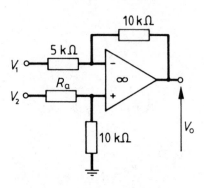

Figure 3.8

(a) Determine the value of R_a necessary to give the stated relationship.
(b) If $V_2 = 10$ V and $V_1 = -10$ V, calculate the current in each resistor.
[*Ans*. 80 kΩ, 1.78 mA, 1.22 mA, 0.11 mA]

3.2. Obtain the expression $\dfrac{V_o}{V_i} = \dfrac{ab}{1 + b \left\{ \dfrac{(1 + a)}{(1 + c)} \right\}}$ for the circuit shown in figure 3.9.

3.3. Derive the expression $\dfrac{V_o}{V_i} = \dfrac{G_3 (G_1 + G_2)}{G_2 G_4 \left\{ \dfrac{G_3 + G_5}{G_4} - \dfrac{G_1}{G_2} \right\}}$ for the circuit shown in figure 3.10 where $G_1 = \dfrac{1}{R_1}$ etc.

If now all of the components, with the exception of G_4, are fixed resistors of equal value, sketch the variation of V_o/V_i with variation of G_4, assuming that it is a pure resistor.

3.4. The circuit shown in figure 3.11 is that of a summing amplifier with inverting gain. The op-amp has an open-loop gain given by $A(j\omega) = A_0 \omega_a / j\omega + \omega_a$. Derive the expression

$$V_o = - \frac{A_0 \omega_a (V_1 + V_2)}{(1 + 2k) \left(j\omega + \omega_a + \dfrac{A_0 \omega_a}{1 + 2k} \right)}$$

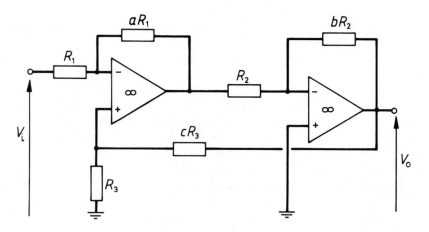

Figure 3.9

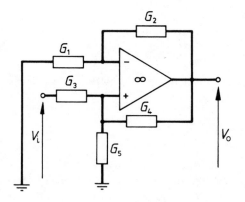

Figure 3.10

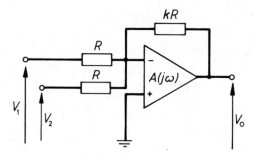

Figure 3.11

Show further that the 3 dB bandwidth is given by $\omega_{3\text{dB}} = A_0\omega_a/1 + 2k$, stating any assumptions made.

3.5. The circuit shown in figure 3.12 is a level-shift network necessary to convert the differential output from the multiplier into a single-ended output relative to earth. Show that the output is given by $V_0 = R_L(I_2 - I_1)$.

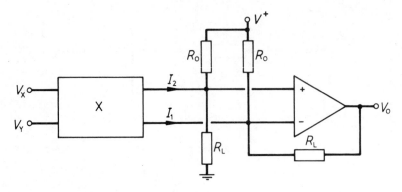

Figure 3.12

3.6. The circuit shown in figure 3.13 is that of an all-pass filter. Derive the relationship

$$\frac{V_0}{V_i} = \frac{A}{2}\left[\frac{j\omega - 1/CR}{j\omega + 1/CR}\right]$$

and obtain the magnitude and phase expressions.

It is now required to simulate a three-phase system of voltages having a frequency of 50 Hz. Design a circuit to achieve the above using the all-pass design assuming ideal op-amps and 1 μF capacitors.

[*Ans.* Two cascaded circuits: $R = 1.84$ kΩ, $C = 1$ μF]

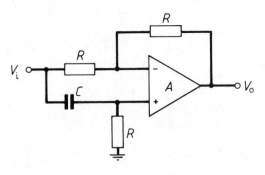

Figure 3.13

4 Basic Filter Circuits

4.1 Introduction

A number of popular circuits will now be considered using the non-inverting (positive gain) and inverting (negative gain) operational amplifier configurations. In both cases, the open-loop gain of the amplifier will be taken to be high (10^5 or greater) and in the first class of circuits to be considered, the so-called voltage-controlled-voltage-source (VCVS), it will be seen that the circuit gain may be adjusted by means of a resistor ratio. In the second class of circuits where there are two feedback paths, the so-called multiple feedback (MFB) circuits, the operational amplifier is again assumed to be working as an ideal *infinite* gain device.

Both circuits have good circuit isolation properties (high input impedance and low output impedance) which means that they may be cascaded to form higher-order filters without the need for additional isolation amplifiers. Both circuits allow for easy design and implementation with a minimum number of components.

Before undertaking the design of the VCVS and MFB second-order circuits, we will consider the analysis and design of the first-order low-pass VCVS filter.

4.2 The voltage-controlled-voltage-source first-order filter

The simplest active filter employing the non-inverting (positive gain) feedback amplifier is shown in figure 4.1. The amplifier is assumed to be a VCVS having a very high open-loop gain A, the circuit gain being controlled by resistors R_3 and R_2 [$K = 1 + (R_3/R_2)$]. Filtering action is performed by the R-C frequency selective network which is cascaded with the amplifier.

Nodal analysis at nodes (a) and (b) yields:

Node (a)

$$\frac{V_a - V_i}{R_1} + j\omega C_1 V_a = 0 \quad \text{or}$$

$$V_a \left(\frac{1}{R_1} + j\omega C_1 \right) = V_i/R_1 \tag{4.1}$$

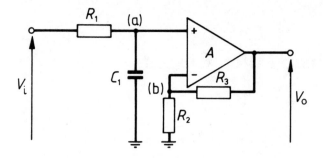

Figure 4.1   First-order low-pass VCVS filter

Node (b)

$$\frac{V_b - V_o}{R_3} + \frac{V_b}{R_2} = 0 \quad \text{or}$$

$$V_b\left(\frac{1}{R_2} + \frac{1}{R_3}\right) = \frac{V_o}{R_3} \quad \text{giving}$$

$$V_b = \left(\frac{R_2}{R_2 + R_3}\right)V_o \tag{4.2}$$

and $V_o = A(V_a - V_b)$, $(V_a - V_b) = \dfrac{V_o}{A} \to 0$ for A large.

Therefore $V_a \approx V_b$ and combining equations (4.1) and (4.2) gives

$$\frac{V_o}{V_i}(j\omega) = \frac{K(1/C_1 R_1)}{j\omega + 1/C_1 R} \tag{4.3}$$

where $K = 1 + R_3/R_2$.

This may be compared with the form for a low-pass circuit:

$$\frac{V_o}{V_i}(j\omega) = \frac{K b_0 \omega_0}{j\omega + b_0 \omega_0} \tag{4.4}$$

Comparing equations (4.3) and (4.4) provides $b_0 \omega_0 = 1/C_1 R_1$, $K = 1 + R_3/R_2$. We see that we have two equations and four unknowns. We therefore select the capacitor value (we have a more restricted choice in capacitors than with resistors).

Example 4.1

A circuit is required to have a voltage gain of 4 in the pass band, to be 3 dB down up to the cut-off frequency of 100 Hz and to be at least 20 dB down at 1 kHz. The design is to be Butterworth using a VCVS circuit.

The performance requirement is shown in figure 4.2.

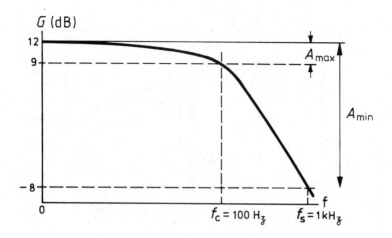

Figure 4.2   Response for example 4.1

To find the order n of the circuit we use equation (2.8):

$$n = \frac{1}{2\log_{10}\left(\dfrac{\omega_s}{\omega_c}\right)} \log_{10}\left\{ \frac{10^{A\,min/10} - 1}{10^{A\,max/10} - 1} \right\}$$

$$n = \frac{1}{2\log_{10}10} \log_{10}\left\{ \frac{10^2 - 1}{10^{0.3} - 1} \right\}$$

$$= 0.5\log_{10}\frac{99}{0.995} = 0.999 \quad \text{or} \quad n = 1$$

Practical experience shows that capacitor values, appropriate to the frequency range under consideration, may be selected by using the following expression:

$$C_1 = \frac{10}{f_c}\,\mu F = \frac{10}{100} \times 10^{-6} = 0.1\,\mu F$$

For a Butterworth response, $b_0 = 1$, therefore

$$R_1 = \frac{1}{b_0\omega_0 C_1} = \frac{1}{2\pi \times 100 \times 10^{-7}} = 15.9\,k\Omega$$

To work out resistors R_2 and R_3, good circuit design suggests that the condition for minimum dc offset should be met. Re-draw the circuit for the dc condition as shown in figure 4.3.

For $V_a = V_b$, $V' = 0$, then $V_{R_1} = V(R_2//R_3)$ or

$$R_1 = \frac{R_2 R_3}{R_2 + R_3} = \frac{R_3}{1 + R_3/R_2} = \frac{R_3}{K}$$

Active Filter Design

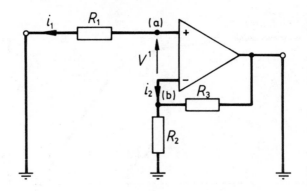

Figure 4.3   DC condition for example 4.1

giving $R_3 = KR_1$.

$$\frac{R_3}{R_2} = K - 1, \text{hence } R_2 = \left(\frac{K}{K-1} \right) R_1$$

Summary of design equations

$$
\boxed{
\begin{array}{ll}
C_1 = \dfrac{10}{f_c}\ \mu\text{F}, & R_2 = \left(\dfrac{K}{K-1} \right) R_1 \\[3mm]
R_1 = \dfrac{1}{b_0 \omega_0 C_1} & R_3 = KR_1
\end{array}
}
$$

To complete the design we find that

$R_2 = 21.2 \text{ k}\Omega$

$R_3 = 63.7 \text{ k}\Omega$

The circuit and response are shown in figure 4.4.

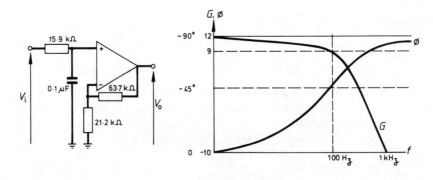

Figure 4.4   Circuit and response for example 4.1

It can be seen that the resistor values are quite large because of the chosen value for the capacitor. Of course the choice of the capacitor value is left to the designer; in the example, selection of capacitor values of 1 μF or 0.47 μF could have been made. Selection of these capacitor values would have resulted in correspondingly reduced values of resistance. This is left as a simple exercise for the reader to perform.

4.2.1 Gain and phase response

From equation (4.4):

$$\left|\frac{V_o}{V_i}(j\omega)\right| = \frac{Kb_0\omega_0}{((b_0\omega_0)^2 + \omega^2)^{0.5}} \; ; \; \phi(j\omega) = -\tan^{-1}\frac{\omega}{b_0\omega_0}$$

For $\omega = \omega_0$ $\quad \left|\frac{V_o}{V_i}(j0)\right| = K; \; \phi(j0) = 0°$

For $\omega = b_0\omega_0$ $\quad \left|\frac{V_o}{V_i}(jb_0\omega_0)\right| = \frac{K}{b_0}$; $\phi(jb_0\omega_0) = -45°$

For $\omega \to \infty$ $\quad \left|\frac{V_o}{V_i}(j\infty)\right| = 0; \; \phi(j\infty) = -90°$

For $\omega \gg \omega_0$, roll-off (dB) = $20\log_{10}\left|\frac{V_o}{V_i}(j\omega)\right| = -20\log_{10}\omega$

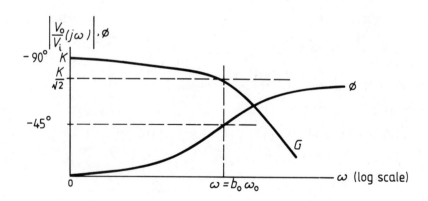

Figure 4.5 Gain/phase responses for first-order VCVS filter

4.3 The voltage-controlled-voltage-source second-order filter

This type of circuit was first proposed by Sallen and Key and is relatively easy to design and implement. It employs the infinite gain VCVS concept associated with the operational amplifier and the circuit offers external gain adjustment via components Y_5 and Y_6. It will be seen in chapter 8 that this type of filter structure, employing the non-inverting (positive gain) operational amplifier, exhibits a high Q-factor sensitivity with the circuit gain $[K = 1 + (R_6/R_5)]$. Because of this high sensitivity, the configuration is restricted to low Q and low gain values with a trade-off being a wide frequency bandwidth.

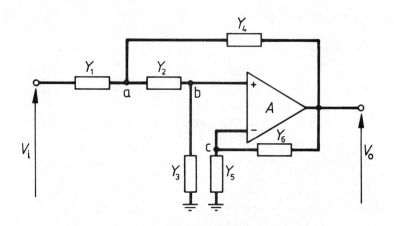

Figure 4.6   Basic second-order VCVS filter

A is assumed very large which implies that $V_b \approx V_c$.

$$V_o = A(V_b - V_c)$$

$$\frac{V_o}{A} = V_b - V_c \to 0 \quad \text{or} \quad V_b \approx V_c$$

Writing nodal equations at (a), (b) and (c) yields:

Node (a)

$$(V_a - V_i)Y_1 + (V_a - V_b)Y_2 + (V_a - V_o)Y_4 = 0$$

$$V_a(Y_1 + Y_2 + Y_4) - V_iY_1 - V_bY_2 - V_oY_4 = 0 \qquad (4.5)$$

Node (b)

$$V_bY_3 + (V_b - V_a)Y_2 = 0$$

$$V_b(Y_3 + Y_2) - V_aY_2 = 0 \qquad (4.6)$$

Node (c)

$$V_cY_5 + (V_c - V_o)Y_6 = 0$$

$$V_c(Y_5 + Y_6) - V_oY_6 = 0 \qquad (4.7)$$

From which

$$V_c = \frac{Y_6}{Y_5 + Y_6} \, V_o \tag{4.8}$$

$$V_b = \frac{Y_2}{Y_2 + Y_3} \, V_a \tag{4.9}$$

$$V_a = \frac{Y_1}{Y_1 + Y_2 + Y_4} \, V_i + \frac{Y_2}{Y_1 + Y_2 + Y_4} \, V_b + \frac{Y_4}{Y_1 + Y_2 + Y_4} \, V_o$$

Substituting this into equation (4.9) yields

$$V_b = \left(\frac{Y_2}{Y_2 + Y_3}\right) \frac{1}{Y_1 + Y_2 + Y_4} \, [Y_1 V_i + Y_2 V_b + Y_4 V_o] \tag{4.10}$$

from which

$$(Y_1 + Y_2 + Y_4) \left(\frac{Y_2 + Y_3}{Y_2}\right) V_b - Y_2 V_b = Y_1 V_i + Y_4 V_o$$

that is

$$V_b \,[(Y_2 + Y_3)(Y_1 + Y_2 + Y_4) - Y_2^2] = Y_1 Y_2 V_i + Y_2 Y_4 V_o \tag{4.11}$$

Equating equations (4.8) and (4.11) yields

$$\frac{Y_6}{Y_5 + Y_6} \, V_o = \frac{Y_1 Y_2 V_i + Y_2 Y_4 V_o}{(Y_2 + Y_3)(Y_1 + Y_2 + Y_4) - Y_2^2} \quad \text{and}$$

$$V_o \left[\frac{Y_6}{Y_5 + Y_6} - \frac{Y_2 Y_4}{(Y_2 + Y_3)(Y_1 + Y_2 + Y_4) - Y_2^2}\right]$$

$$= \frac{Y_1 Y_2 V_i}{(Y_2 + Y_3)(Y_1 + Y_2 + Y_4) - Y_2^2}$$

Finally, after some manipulation, we obtain the expression

$$\frac{V_o}{V_i} = \frac{Y_1 Y_2 \, (Y_5 + Y_6)/Y_6}{[(Y_2 + Y_3)(Y_1 + Y_2 + Y_4) - Y_2^2] - Y_2 Y_4 \, (Y_5 + Y_6)/Y_6} \tag{4.12}$$

We will now consider the several filter types which the circuit will yield for a given association of components.

4.3.1 The low-pass filter

Here $Y_1 = \frac{1}{R_1}$, $Y_2 = \frac{1}{R_2}$, $Y_3 = j\omega C_1$, $Y_4 = j\omega C_2$, $Y_5 = \frac{1}{R_3}$, $Y_6 = \frac{1}{R_4}$.

Equation (4.12) yields the following expression upon insertion of the components:

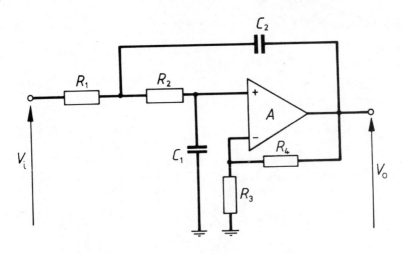

Figure 4.7   Second-order low-pass VCVS filter

$$\frac{V_o}{V_i}(j\omega) = \cfrac{\dfrac{1}{R_1 R_2}\left(\dfrac{1}{R_3} + \dfrac{1}{R_4}\right)\Big/\dfrac{1}{R_4}}{\left(\dfrac{1}{R_2} + j\omega C_1\right)\left(\dfrac{1}{R_1} + \dfrac{1}{R_2} + j\omega C_2\right) - \dfrac{1}{R_2^2} - \dfrac{j\omega C_2}{R_2}\left(\dfrac{1}{R_3} + \dfrac{1}{R_4}\right)\Big/\dfrac{1}{R_4}}$$

Manipulation of the expression yields

$$\frac{V_o}{V_i}(j\omega) = \cfrac{\dfrac{1}{R_1 R_2}\left(1 + \dfrac{R_4}{R_3}\right)}{-\omega^2 C_1 C_2 + \dfrac{j\omega C_1}{R_1} + \dfrac{j\omega C_1}{R_2} + \dfrac{j\omega C_2}{R_2} - \dfrac{j\omega C_2}{R_2}\left(1 + \dfrac{R_4}{R_3}\right) + \dfrac{1}{R_1 R_2}}$$

Now $K = 1 + R_4/R_3$ which is the dc gain of the stage and the expression may finally be written as

$$\frac{V_o}{V_i}(j\omega) = \cfrac{\dfrac{1}{C_1 C_2 R_1 R_2}K}{-\omega^2 + j\omega\left\{\dfrac{1}{C_2 R_1} + \dfrac{1}{C_2 R_2} + \dfrac{1}{C_1 R_2} - \dfrac{K}{C_1 R_2}\right\} + \dfrac{1}{C_1 C_2 R_1 R_2}} \tag{4.13}$$

If we note further that $\omega_0^2 = \dfrac{1}{C_1 C_2 R_1 R_2}$ then we may compare equation (4.13) with the standard form for a second-order low-pass filter:

$$\frac{V_o}{V_i}(j\omega) = \frac{K b_0}{-\omega^2 + j b_1 \omega \omega_0 + b_0 \omega_0^2} \tag{4.14}$$

where ω_0 is the 3 dB frequency for the Butterworth case and $b_0\omega_0$ is the frequency at the end of the ripple band for the Chebyshev filter. The coefficients b_0, b_1 etc. are the appropriate coefficients for the prescribed filter response.

Comparison between equations (4.13) and (4.14) yields

$$\left.\begin{aligned} b_0\omega_0^2 &= \frac{1}{C_1 C_2 R_1 R_2} \\ b_1\omega_0 &= \frac{1}{C_2 R_1} + \frac{1}{C_2 R_2} + \frac{1}{C_1 R_2} + \frac{K}{C_1 R_2} \end{aligned}\right\} \qquad (4.15)$$

We must now introduce normalisation into the components and design parameters in order to produce a rationalised design procedure. Let $\omega_0^2 = 1$ rad/s, $C_1 = C_2 = 1$ F. Then equation (4.15) becomes

$$\left.\begin{aligned} b_0 &= \frac{1}{R_1 R_2} \\ b_1 &= \frac{1}{R_1} + \frac{1}{R_2}(2 - K) \end{aligned}\right\} \qquad (4.16)$$

Notice that there are six unknowns and therefore a certain degree of component selectivity will have to be made since a unique solution is not possible.

From equation (4.16), $R_1 = 1/b_0 R_2$ and on substituting into the second equation (4.16) we obtain

$$b_1 = b_0 R_2 + \frac{(2 - K)}{R_2}$$

which may be written

$$b_0 R_2^2 - b_1 R_2 + (2 - K) = 0$$

and which has a solution

$$R_2 = (b_1 \pm \sqrt{(b_1^2 - 4b_0(2 - K))})/2b_0 \qquad (4.17)$$

So far we have normalised the capacitors C_1, C_2 and obtained the normalised values for the resistors R_1, R_2. We now have to obtain values for the gain resistors R_3, R_4 for the good circuit design condition of minimum dc offset.

Figure 4.8 shows the re-drawn circuit for the dc condition where the capacitors are now represented by open circuits.

For minimum dc offset, $V_+ = V_-$ or

$$V(R_1 + R_2) = V(R_3//R_4)$$

hence

$$R_1 + R_2 = \frac{R_3 R_4}{R_3 + R_4} = \frac{R_4/R_3}{1 + R_4/R_3}$$

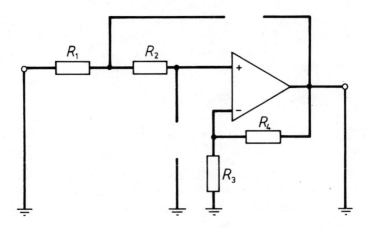

Figure 4.8   DC condition for second-order low-pass filter

giving

$$\frac{R_4}{R_3} = K - 1 = K(R_1 + R_2)$$

hence

$$'R_3 = \left(\frac{K}{K-1}\right)(R_1 + R_2); \; R_4 = K(R_1 + R_2)$$

We now have all the necessary information to enable a complete design to be achieved.

Example 4.2

A low-pass filter is required to provide the specifications given in figure 4.9. The design is to be a Butterworth maximally flat realisation having a pass-band gain of 20 dB.

The requirement is for the output to be attenuated by 3 dB at 2 kHz and to have at least 40 dB of attenuation at 20 kHz.

Using equations from chapter 2 we obtain the order n of the filter and its performance in the pass and stop bands:

$$n = \frac{1}{2\log_{10}\left(\dfrac{\omega_s}{\omega_c}\right)} \; \log_{10} \left\{ \frac{10^{A\,\text{min}/10} - 1}{10^{A\,\text{max}/10} - 1} \right\}$$

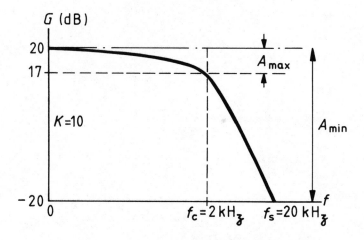

Figure 4.9   Specification for example 4.2

Insert values:

$$n = \frac{1}{2\log_{10}10} \quad \log_{10}\left\{\frac{10^4 - 1}{10^{0.3} - 1}\right\}$$

producing $n = 2$ (a second-order circuit). At 2 kHz

$$A_{max} = 10\log_{10}\left(1 + \left(\frac{\omega}{\omega_c}\right)^{2n}\right)$$

$$= 10\log_{10}(1 + 1^4)$$

$$= 3 \text{ dB for } \omega = \omega_c$$

$$A_{min} = 10\log_{10}\left(1 + \left(\frac{20}{2}\right)^4\right) = 40 \text{ dB for } \omega = \omega_s$$

The filter will therefore perform exactly as required by the design.

We now choose values for the capacitors and a good choice is made using the formula $C = (10/f_c)$ μF. Here $f_c = 2$ kHz giving $C_1 = C_2 = 10/(2 \times 10^3) = 5$ nF. The Butterworth coefficients for $n = 2$ are found from the table 2.1: $b_0 = 1$, $b_1 = 1.414$, hence

$$R_2 = \frac{1.414 \pm \sqrt{[2 - 4(2 - 10)]}}{2} = \frac{1.414 \pm 5.83}{2}$$

$$= 3.622 \ \Omega \text{ (the negative value is inadmissible) and}$$

$$R_1 = \frac{1}{b_0R_2} = 0.276 \ \Omega$$

$$R_3 = \frac{10}{9} \times 3.898 = 4.33 \ \Omega$$

$$R_4 = 10 \times 3.898 = 38.98 \ \Omega$$

Scaling factors:

$$K_f = \frac{\omega_o}{\omega_n} = \frac{2\pi \times 2 \times 10^3}{1} = 4\pi \times 10^3$$

$$k_m k_f = \frac{C_n}{C_o} \text{ giving } k_m = \frac{1}{4\pi \times 10^3 \times 5 \times 10^{-9}} = 1.59 \times 10^4$$

Scaled values:

$$R_1 = 0.276 \times 1.59 \times 10^3 = 4.39 \ k\Omega; C_1 = C_2 = 5 \ nF$$

$$R_2 = 3.622 \times 1.59 \times 10^3 = 57.5 \ k\Omega$$

$$R_3 = 4.33 \times 1.59 \times 10^3 = 68.9 \ k\Omega$$

$$R_4 = 38.98 \times 1.59 \times 10^3 = 620.3 \ k\Omega$$

The circuit is shown in figure 4.10 with its response.

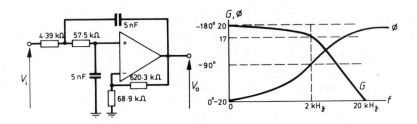

Figure 4.10   Final circuit for example 4.2

It should be pointed out that an adjustment to the scaling may be made in order that realistic values of resistors and capacitors may be achieved. Consider that we select $C_1 = C_2 = 0.1 \ \mu F$, then

$$k_m = \frac{C_n}{k_f C_o} = \frac{1}{4\pi \times 10^3 \times 10^{-7}} = 795$$

The resistor values would be

$$R_1 = 219.6 \ \Omega$$
$$R_2 = 2.88 \ k\Omega$$
$$R_3 = 3.45 \ k\Omega$$
$$R_4 = 31 \ k\Omega$$
$$C_1 = C_2 = 0.1 \ \mu F$$

It can be seen that with a larger value of capacitor, the resistor values are more realistic.

4.3.2 Gain and phase shift of the circuit

Re-call equation (4.14) from which

$$\left. \begin{array}{c} \left| \dfrac{V_o}{V_i} (j\omega) \right| = \dfrac{K b_0 \omega_0^2}{[(b_0 \omega_0^2 - \omega^2)^2 + (b_1 \omega \omega_0)^2]^{0.5}} \\[4mm] \phi(j\omega) = -\tan^{-1} \dfrac{(b_1 \omega \omega_0)}{(b_0 \omega_0^2 - \omega^2)} \end{array} \right\} \tag{4.18}$$

A consideration of the three frequencies used in the case of the VCVS circuit yields

$$\omega = 0 \qquad \left| \frac{V_o}{V_i} (j0) \right| = K, \ \phi(j0) = 0°$$

$$\omega = \infty \qquad \left| \frac{V_o}{V_i} (j\infty) \right| \to 0, \ \phi(j\infty) \to -180°$$

$$\omega' = \sqrt{b_0}\,\omega_0 \qquad \left| \frac{V_o}{V_i} (j\omega') \right| = \frac{K\sqrt{b_0}}{b_1}, \ \phi(j\omega') = -90°$$

These are shown in figure 4.11 for $b_0 = 1, b_1 = \sqrt{2}$

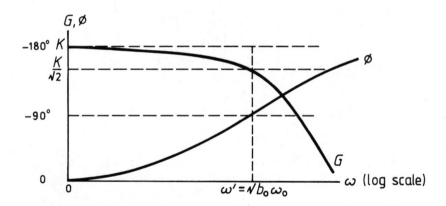

Figure 4.11 Gain/phase responses for second-order VCVS filter

An insight into the action of the filter may be obtained by noting that at low frequencies the reactances of C_1 and C_2 are high and the signal is applied via resistors R_1 and R_2 with negligible attenuation. At higher frequencies the reactances of C_1 and C_2 are lowered significantly. Capacitor C_1 shunts a propor-tion of the input signal to ground, thereby reducing the potential at node b, while C_2 feeds back a proportion of the output to node a with a corresponding rise in nodal potential. There are two roll-off paths, via $(R_1\,C_2)$ and $(R_2\,C_1)$ which combine to produce the second-order response.

Active Filter Design

4.4 The VCVS high-pass filter

This circuit is shown in figure 4.12 from which note that the frequency selective
components (R_1, R_2, C_1, C_2) are interchanged with those of the low-pass filter
shown in figure 4.7. Notice that the resistors *become* capacitors and capacitors
become resistors. This concept is associated with what is referred to as *frequency
transformation* and is useful in translating a high-pass filter specification into an
equivalent low-pass specification. The low-pass design (prototype) is realised as
in example 4.2, after which frequency transformation is applied to realise the
prototype high-pass filter. Finally, frequency and magnitude scaling produces
the final circuit design.

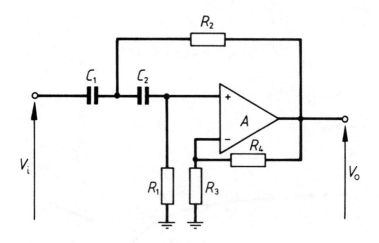

Figure 4.12   Second-order high-pass VCVS filter

For the high-pass filter the components are $Y_1 = j\omega C_1$, $Y_2 = j\omega C_2$, $Y_3 = 1/R_1$,
$Y_4 = 1/R_2$, $Y_5 = 1/R_3$, $Y_6 = 1/R_4$. Insertion into (4.12) yields the expression:

$$\frac{V_o}{V_i}(j\omega) = \frac{-\omega^2 C_1 C_2 (1 + R_4/R_3)}{\left(j\omega C_2 + \dfrac{1}{R_1}\right)\left(j\omega C_1 + j\omega C_2 + \dfrac{1}{R_2}\right) + \omega^2 C_2^2 - \dfrac{j\omega C_2}{R_2}\left(1 + \dfrac{R_4}{R_3}\right)}$$

and

$$\frac{V_o}{V_i}(j\omega) = \frac{-\omega^2 C_1 C_2 K}{-\omega^2 C_1 C_2 - \omega^2 C_2^2 + \dfrac{j\omega C_2}{R_2} + \omega^2 C_2^2 + \dfrac{j\omega C_1}{R_1} + \dfrac{j\omega C_2}{R_1} + \dfrac{1}{R_1 R_2} - \dfrac{j\omega C_2 K}{R_2}}$$

which on simplification becomes

$$\frac{V_o}{V_i}(j\omega) = \frac{-\omega^2 K}{-\omega^2 + j\omega\left(\dfrac{1}{C_1 R_1} + \dfrac{1}{C_2 R_1} + \dfrac{(1-K)}{C_1 R_2}\right) + \dfrac{1}{C_1 C_2 R_1 R_2}} \tag{4.19}$$

Equation (4.19) should be compared with the standard form for a high-pass filter given by

$$\frac{V_o}{V_i}(j\omega) = \frac{-\omega^2 K}{-\omega^2 + j\omega\omega_0 B_1 + B_0 \omega_0^2} \tag{4.20}$$

from which we obtain, noting that $B_0 = 1/b_0, B_1 = b_1/b_0$:

$$B_0 \omega_0^2 = \frac{1}{C_1 C_2 R_1 R_2}$$

$$B_1 \omega_0 = \frac{(1-K)}{C_1 R_2} + \frac{1}{C_1 R_1} + \frac{1}{C_2 R_1}$$

Again normalising ω_0, C_1, C_2 as before, $\omega_0 = 1$ rad/s, $C_1 = C_2 = 1$ F. Therefore

$$B_0 = \frac{1}{R_1 R_2} \quad \text{and} \quad B_0 R_2 = \frac{1}{R_1}$$

$$B_1 = \frac{(1-K)}{R_2} + 2B_0 R_2$$

from which is obtained

$$R_2 = \frac{[-B_1 \pm (B_1^2 - 8B_0(1-K))^{0.5}]}{4B_0} \tag{4.21}$$

The method outlined for obtaining the high-pass design in section 4.4 is the recommended design procedure to follow, since the low-pass coefficients b_0, b_1 are readily available from printed tables of data. It is relatively rare to find printed tables of the high-pass coefficients B_0, B_1.

Example 4.3

A high-pass filter is required to meet the specifications shown in figure 4.13. Obtain the component values necessary to satisfy the requirement.

The response will be Chebyshev having a ripple width of 3 dB and a high-frequency gain of 20 dB. The low-pass coefficients for $A_{max} = 3$ dB are from table 2.4: $b_1 = 0.65, b_0 = 0.708$. Using equation (2.15) to determine the order n of the filter and noting that for the identical 3 dB frequency the prototype low-pass filter will have the inverse of the ratio of frequencies as is the case for the high-pass filter:

Active Filter Design

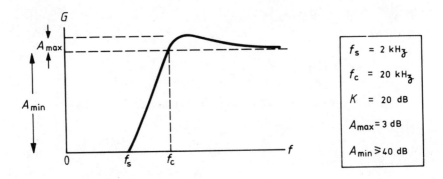

Figure 4.13   Response characteristic for example 4.3

$$\left.\frac{\omega_s}{\omega_c}\right|_{LP} = \left.\frac{\Omega_c}{\Omega_s}\right|_{HP} \tag{4.22}$$

Therefore

$$\frac{\omega_s}{\omega_c} = \frac{20}{2} = 10 \quad \text{and} \quad \omega_s = 10 \times \omega_c = 200 \text{ kHz}$$

also the order

$$n = \cosh^{-1} \frac{[(10^4 - 1)/(10^{0.3} - 1)]^{0.5}}{\cosh^{-1} 10} = 1.77 \text{ (round up to 2)}$$

Figure 4.14 shows the high-pass characteristic frequency-translated into the equivalent low-pass specification.

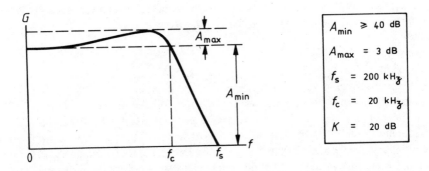

Figure 4.14   Low-pass response characteristic for example 4.3

The value for n suggests that the attenuation in the stop band will be greater than that for the identical order Butterworth filter. Using equation (2.14) and noting that from equation (2.6) $\epsilon = (10^{0.3} - 1)^{0.5} = 1$:

$$A_{min} = 10\log_{10} [1 + (\cosh 2 \cosh^{-1} 10)^2] = 45.9 \text{ dB}$$

The prototype *normalised* low-pass circuit may now be designed.
Let $C_1 = C_2 = 1$ F, $\omega_0 = 1$ rad/s.
Using equations (4.16) and (4.17)

$$R_1 = \frac{1}{b_0 R_2}, \quad R_2 = \frac{[b_1 \pm (b_1^2 - 4b_0(2 - K))^{0.5}]}{2b_0}$$

Also noting for $\epsilon = 1$, $A_{max} = 3$ dB and from table 2.4 $b_0 = 0.708$, $b_1 = 0.65$:

$$R_2 = \frac{0.65 \pm (0.65^2 - 4 \times 0.708 (2 - 10))^{0.5}}{2 \times 0.708} = 3.85 \ \Omega$$

$$R_1 = \frac{1}{0.708 \times 3.85} = 0.37 \ \Omega$$

$$R_3 = \frac{10}{9 \times 4.22} = 4.7 \ \Omega$$

$$R_4 = 10 \times 4.22 = 42.2 \ \Omega$$

$$C_1 = C_2 = 1 \text{ F}$$

We must now apply the frequency transformation to the low-pass components in order to obtain the equivalent high-pass components. A brief outline of the concept is now included. We have seen that there is an inverse relationship between the ratio of the stop-band and pass-band frequencies for the low- and high-pass circuits. In general this may be written (for a normalised cut-off frequency)

$$\Omega \text{ (HP)} = \frac{1}{\omega} \text{ (LP)}$$

which for the VCVS low-pass $\omega_0 = \dfrac{1}{(b_0 C_1 C_2 R_1 R_2)^{0.5}}$ and

$$\Omega_0 = (b_0 C_1 C_2 R_1 R_2)^{0.5} = \left(\frac{b_0}{R_1' R_2' C_1' C_2'}\right)^{0.5}$$

where $R_1' = \dfrac{1}{C_1}$

$$R_2' = \frac{1}{C_2}$$

$$C_1' = \frac{1}{R_1}$$

$$C_2' = \frac{1}{R_2}$$

where it can be seen that the capacitor (LP) becomes a resistor (HP) and a resistor (LP) becomes a capacitor (HP).

Therefore, for the prototype high-pass the normalised components may be calculated:

$$C_1 = \frac{1}{0.37} = 2.702 \text{ F}, \ C_2 = \frac{1}{3.85} = 0.26 \text{ F}$$

$$R_1 = \frac{1}{1} = 1 \ \Omega, \ R_2 = \frac{1}{1} = 1 \ \Omega, \ R_3 = 4.7 \ \Omega, \ R_4 = 42.4 \ \Omega$$

The gain resistors are unaffected by the transformation. For scaling, select $C = 0.01 \ \mu\text{F}$ which yields

$$k_f = 2\pi \times 2 \times 10^4 = 4\pi \times 10^4 \qquad k_m = \frac{10^8}{4\pi \times 10^4} = 796$$

The components are

$$R_1 = R_2 = 796 \ \Omega \quad R_3 = 3.74 \text{ k}\Omega \quad R_4 = 33.6 \text{ k}\Omega$$

$$C_1 = 2.702/k_m k_f = 0.027 \ \mu\text{F}$$

$$C_2 = 0.26/k_m k_f = 0.0026 \ \mu\text{F}$$

The final circuit is shown in figure 4.15 with a CAD prediction of the response. Note that the program gives $\phi(j\omega)$ from $-180°$ to $-360°$.

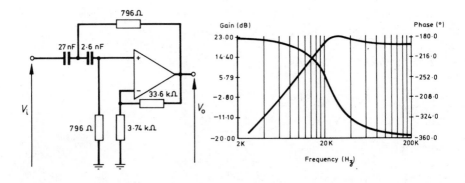

Figure 4.15   Circuit and gain/phase responses for VCVS high-pass filter

4.4.1 Gain and phase response

From equation (4.20):

$$\left| \frac{V_o}{V_i} (j\omega) \right| = \frac{\omega^2 K}{\{(b_0\omega_0^2 - \omega^2) + (\omega\omega_0 b_1)^2\}^{0.5}}$$

$$\phi(j\omega) = 180° - \tan^{-1} \frac{\omega\omega_0 b_1}{b_0\omega_0^2 - \omega^2}$$

Insertion of $\omega = 0$, $\omega = \sqrt{b_0}\,\omega_0$, $\omega = \infty$ yields

$$\left| \frac{V_o}{V_i} (j\omega) \right| = 0, \frac{\sqrt{b_0}K}{b_1}, \ K: \phi(j\omega) = 180°, 90°, 0°$$

The predicted shape of the response is seen from figure 4.15.

Circuit operation is clearly the converse of that for the low-pass filter, as explained earlier. The reactances of C_1 and C_2 are very high at low frequencies, thereby preventing the signal from reaching the op-amp. At higher frequencies these reactances are very low and the signal may pass with negligible attenuation. Feedback increases via resistor R_2 and there are again two roll-off paths.

4.5 The VCVS band-pass filter

The basic structure of the circuit is that shown in figure 4.16 where the components are

$$Y_1 = \frac{1}{R_1}, \ Y_2 = j\omega C_1, \ Y_3 = \frac{1}{R_3}, \ Y_4 = \frac{1}{R_2}, \ Y_5 = \frac{1}{R_5}$$

$$Y_6 = \frac{1}{R_4}, \ Y_7 = j\omega C_2$$

Notice that a capacitor C_2 is connected between node a and ground.

Nodal analysis yields the three equations

$$V_a \left(\frac{1}{R_1} + \frac{1}{R_2} + j\omega(C_1 + C_2) \right) = \frac{V_i}{R_1} + j\omega C_1 V_b + \frac{V_o}{R_2}$$

$$V_b \left(\frac{1}{R_3} + j\omega C_1 \right) = j\omega C_1 V_a \tag{4.23}$$

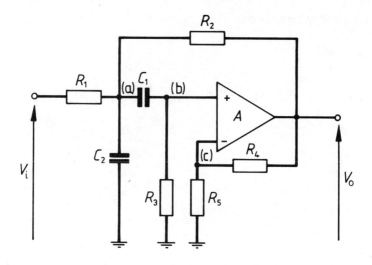

Figure 4.16   VCVS band-pass circuit

$$V_c = \frac{V_o}{K} \text{ , where}$$

$$K = 1 + \frac{R_4}{R_5}$$

Also for A large, $V_b \approx V_c$ and

$$\frac{V_o}{V_i} (j\omega) = \frac{j\omega K/C_2 R_1}{-\omega^2 + \dfrac{j\omega}{C_2}\left[\dfrac{1}{R_1} + \dfrac{1}{R_3} + \dfrac{1}{R_2}(1-K) + \dfrac{C_2}{C_1 R_3}\right] + \dfrac{(R_1+R_2)}{C_1 C_2 R_1 R_2 R_3}}$$

$$(4.24)$$

The general form for a band-pass function having a positive gain is given by the expression

$$\frac{V_o}{V_i} (j\omega) = \frac{j\omega\omega_0 G/Q}{-\omega^2 + j\omega\omega_0/Q + \omega_0^2} \tag{4.25}$$

In the design of band-pass filters, three design parameters must be specified. They are centre frequency gain (G), centre frequency (f_0) and the Q-factor of the circuit (Q). The relationship between the Q-factor and the centre frequency also provides the 3 dB bandwidth of the circuit:

$$\Delta f = f_2 - f_1 = f_0/Q \tag{4.26}$$

These relationships are shown in figure 4.17.

Comparing equations (4.24) and (4.25) yields for normalised values of $C_1 = C_2 = 1$ F, $\omega_0 = 1$ rad/s:

$$G/Q = \frac{K}{R_1} = KG$$

$$1/Q = G_1 + G_2(1 - K) + 2G_3 \tag{4.27}$$

$$G_3 = \frac{1}{G_1 + G_2}$$

where $G_1 = \dfrac{1}{R_1}$ etc.

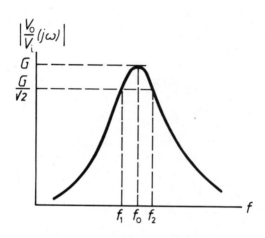

Figure 4.17   Gain response of a band-pass filter

At this point it is expedient to choose the condition that $K = 2$, which implies that $R_4 = R_5$. The reader may pursue conditions other than those for $K = 2$ but will find that the design equations become difficult to manipulate. The equations (4.27) now become

$$G_2 = \left\{ \frac{-1 \pm ((G - 1)^2 + 8Q^2)^{0.5}}{2Q} \right\}$$

$$G_1 = \frac{G}{2Q}$$

$$G_3 = \frac{1}{G_1 + G_2}$$

Example 4.4

A band-pass filter is required to have a centre frequency gain of 20 dB at 1 kHz and a Q-factor of 2. Obtain the design using 0.01 μF capacitors.

Using equation (4.26) the bandwidth is found to be Δf = 1000/2 = 500 Hz, giving f_0 = 1 kHz, f_2 = 1.25 kHz, f_1 = 750 Hz.
Choose K = 2; $R_4 = R_5$:

$$G_2 = \left\{ \frac{-1 \pm ((10-1)^2 + 8 \times 4)^{0.5}}{4} \right\} = 2.408 \text{ s}$$

$$G_1 = \frac{10}{4} = 2.5 \text{ s}$$

$$G_3 = \frac{1}{4.908} = 0.204 \text{ s}$$

$$C_1 = C_2 = 1 \text{ F}$$

Scaling

$$k_f = 2\pi \times 10^3, \quad k_m = \frac{10^8}{2\pi \times 10^3} = 1.59 \times 10^4$$

$$R_1 = 1.59 \times 10^4 \times \frac{1}{2.5} = 6.36 \ \Omega$$

$$R_2 = 1.59 \times 10^4 \times \frac{1}{2.408} = 6.603 \text{ k}\Omega$$

$$R_3 = 1.59 \times 10^4 \times \frac{1}{0.204} = 77.9 \text{ k}\Omega$$

$$C_1 = C_2 = 0.01 \ \mu\text{F}$$

$$R_4 = KR_3 = 155.8 \text{ k}\Omega$$

$$R_5 = \left(\frac{K}{K-1} \right) R_3 = 155.8 \text{ k}\Omega$$

The circuit with its response is shown in figure 4.18.

The operation of the circuit may be summarised briefly as follows. Capacitors C_1 and C_2 have high reactance values at low frequencies and maximum attenuation of the input signal occurs. At high frequencies these reactances are very low and most of the input signal and the feedback signal through R_2 are by-passed to earth. Over a narrow range of frequencies around the centre frequency, frequency selection takes place, involving R_1, R_2, R_3, C_1 and C_2. Note that at low and high frequencies the circuit approximates to a low-pass and then a high-pass filter.

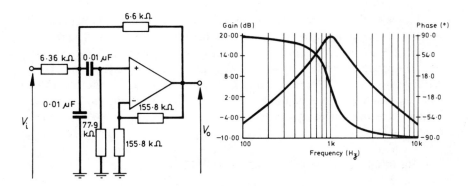

Figure 4.18   Gain/phase responses for VCVS band-pass filter

4.5.1 Gain and phase response

Re-stating equation (4.25):

$$\frac{V_o}{V_i}(j\omega) = \frac{j\omega\omega_0 G/Q}{-\omega^2 + j\omega\omega_0/Q + \omega_0^2} \quad \text{from which}$$

$$\left|\frac{V_o}{V_i}(j\omega)\right| = \frac{\omega\omega_0 G/Q}{\left((\omega_0^2 - \omega^2)^2 + \left(\frac{\omega\omega_0}{Q}\right)^2\right)} \ , \ \phi(j\omega) = 90° - \tan^{-1}\frac{\omega\omega_0}{\omega_0^2 - \omega^2}$$

Insertion of $\omega = 0$, $\omega = \omega_0$, $\omega \to \infty$ yields the response form shown in figure 4.19.

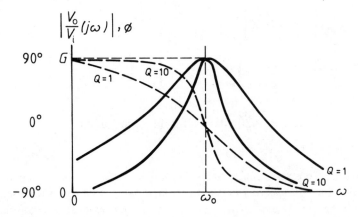

Figure 4.19   Response of VCVS band-pass filter for various values of Q

The general advantages of all the structures considered so far are that they are relatively easy to design and implement. The designer has control over the element values and the *spread* of these values. The disadvantages are those involving high sensitivity to component changes and low Q-factor realisations. The reason for such low Q realisations lies with having a positive gain allied with the roots of the general second-order denominator expression being close to the $(j\omega)$ axis. Obviously for reasons of stability, low-gain and low-Q circuits are chosen. Nevertheless the VCVS circuits are easier to tune than the infinite gain (MFB) counterparts, and are also adjustable over a wide frequency range without affecting the network parameters.

4.6 The infinite gain multiple feedback (MFB) filter

This circuit is an alternative to the previously considered VCVS circuit which had a relatively low closed-loop gain, insofar as the operational amplifier is deemed to be working as an *ideal* VCVS with an infinite gain. The overall circuit gain is of course finite and fixed by the circuit components. The expression *multiple feedback* refers to the two feedback paths through admittances Y_3 and Y_4 as can be seen in figure 4.20, from which it will be seen that the output has a $180°$ phase shift in addition to the phase shift through the network.

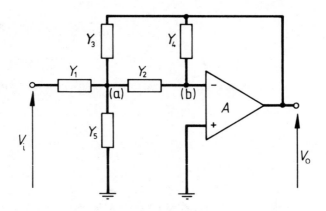

Figure 4.20   Basic MFB filter configuration

This circuit possesses much smaller sensitivities than the VCVS circuit previously considered, especially that of the Q-factor associated with the closed-loop gain. With this configuration it is possible to realise reasonably large values of Q-factor at low frequencies, although the trade-off is a narrower bandwidth and a much higher open-loop gain requirement. Although we are assuming that the gain is infinite when conducting the analysis, it is in reality far from being the case.

Assuming that $A \to \infty$ and $V_b \to 0$, nodal analysis at nodes (a) and (b) gives:

Node (a)

$$(V_a - V_i)Y_1 + V_a Y_5 + (V_a - V_o)Y_3 + (V_a - 0)Y_2 = 0$$

$$V_a(Y_1 + Y_2 + Y_3 + Y_5) - V_i Y_1 - V_o Y_3 = 0 \qquad (4.28)$$

Node (b)

$$(0 - V_a)Y_2 + (0 - V_o)Y_4 = 0$$

$$V_a = -\frac{Y_4}{Y_2} V_o \qquad (4.29)$$

which produces, when inserted into equation (4.28)

$$\frac{V_o}{V_i} = -\frac{Y_1 Y_2}{Y_4(Y_1 + Y_2 + Y_3 + Y_5) + Y_2 Y_3} \qquad (4.30)$$

4.6.1 The multiple feedback low-pass filter

The circuit configuration is shown in figure 4.21 where $Y_1 = 1/R_1$, $Y_2 = 1/R_3$, $Y_3 = 1/R_2$, $Y_4 = j\omega C_2$, $Y_5 = j\omega C_1$.

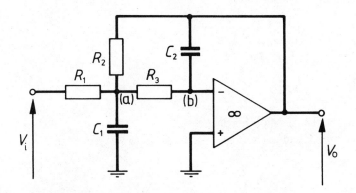

Figure 4.21   MFB low-pass filter

Insertion of admittances into equation (4.30) gives the expression

$$\frac{V_o}{V_i}(j\omega) = \frac{-\dfrac{1}{R_1 R_3}}{j\omega C_2 \left(\dfrac{1}{R_1} + \dfrac{1}{R_3} + \dfrac{1}{R_2} + j\omega C_1\right) + \dfrac{1}{R_1 R_2}}$$

which after manipulation becomes

$$\frac{V_o}{V_i}(j\omega) = -\frac{\dfrac{1}{R_1 R_3 C_1 C_2}}{-\omega^2 + \dfrac{j\omega}{C_1}\left(\dfrac{1}{R_1} + \dfrac{1}{R_2} + \dfrac{1}{R_3}\right) + \dfrac{1}{C_1 C_2 R_2 R_3}} \tag{4.31}$$

This expression may be compared with that for a typical low-pass response, namely

$$\frac{V_o}{V_i}(j\omega) = \frac{-K b_0 \omega_0^2}{-\omega^2 + j b_1 \omega \omega_0 + b_0 \omega_0^2} \tag{4.32}$$

The gain and phase responses have the same shape as for the VCVS filter but it should be noted that the MFB filter provides an *extra* phase shift of 180° because of the input signal being applied to the inverting terminal. Therefore

$$\phi(j\omega) = 180° - \tan^{-1}\frac{b_1 \omega \omega_0}{b_0 \omega_0^2 - \omega^2} \tag{4.33}$$

Comparing equation (4.31) and equation (4.32) yields

$$Kb_0 \omega_0^2 = \frac{1}{R_1 R_3 C_1 C_2}$$

$$b_0 \omega_0^2 = \frac{1}{C_1 C_2 R_2 R_3} \tag{4.34}$$

$$b_1 \omega_0 = \frac{1}{C_1}\left(\frac{1}{R_1} + \frac{1}{R_2} + \frac{1}{R_3}\right)$$

From which the dc gain is $K = R_2/R_1$.

Let $C_1 = 1$ F and $\omega_0 = 1$ rad/s, so equations (4.34) reduce to

$$Kb_0 = \frac{1}{R_1 R_3 C_2} \, , \quad \frac{1}{R_1} = Kb_0 R_3 C_2 = \frac{K}{R_2}$$

$$b_0 = \frac{1}{R_2 R_3 C_2} \, , \quad \frac{1}{R_3} = b_0 R_2 C_2$$

$$b_1 = \left(\frac{1}{R_1} + \frac{1}{R_2} + \frac{1}{R_3}\right)$$

which on substituting for $1/R_1$ and $1/R_3$ gives

$$b_1 = \frac{K}{R_2} + \frac{1}{R_2} + b_0 R_2 C_2$$

which yields the quadratic expression

$$b_0 R_2^2 = \frac{b_1 R_2}{C_2} + \frac{(1+K)}{C_2} = 0 \qquad (4.35)$$

which has a solution

$$R_2 = \frac{\frac{b_1}{C_2} \pm \left(\left(\frac{b_1}{C_2}\right)^2 - \frac{4(1+K)b_0}{C_2}\right)^{0.5}}{2b_0} \qquad (4.36)$$

For R_2 to be real and positive, the following condition for C_2 must prevail;

$$\left(\frac{b_1}{C_2}\right)^2 \geqslant \frac{4(1+K)b_0}{C_2} \quad \text{or}$$

$$b_1^2 \geqslant 4(1+K)C_2 b_0 \qquad \text{and}$$

$$C_2 \leqslant \frac{b_1^2}{4(1+K)b_0} \qquad (4.37)$$

Note here that there are five unknowns and three basic equations, and note again that a unique solution is not possible.

Example 4.5

Consider now the design specifications of the previous example 4.2 which are summarised here:

A_{max}(pass-band) 3 dB, $K = 10$

A_{min}(stop-band) 40 dB, $b_1 = \sqrt{2}, b_0 = 1$

$f_c = 2$ kHz, $f_s = 20$ kHz

The circuit is of the second order as calculated. For normalised $\omega_0 = 1$ rad/s and $C_1 = 1$ F obtain $C_2 \leqslant \frac{2}{4 \times 11} \leqslant 0.045$ F *normalised*. Choose $C_2 = 0.01$ F and

$$R_2 = \frac{\frac{\sqrt{2}}{0.01} \pm \left(\left(\frac{\sqrt{2}}{0.01}\right)^2 - \frac{4(11)}{0.01}\right)^{0.5}}{2}$$

This yields $R_2 = 133\ \Omega$ or $8.3\ \Omega$, giving

$$R_1 = \frac{R_2}{K} = 13.3\ \Omega \text{ or } 0.83\ \Omega$$

$$R_3 = \frac{1}{b_0 R_2 C_2} = 0.752\ \Omega \text{ or } 12.12\ \Omega$$

Select $C_1 = \dfrac{10}{f_c}\ \mu F = \dfrac{10}{2 \times 10^3} = 5\ nF.$

Scaling:

$$k_f = 2\pi \times 2 \times 10^3 = 4\pi \times 10^3$$

$$k_m = \frac{C_n}{k_f C_0} = \frac{1}{4\pi \times 10^3 \times 5 \times 10^{-9}} = 1.59 \times 10^4$$

Then

$R_1 = 2.117\ k\Omega$ or $13.2\ k\Omega$	
$R_2 = 2.11\ M\Omega$ or $132\ k\Omega$	
$R_3 = 11.96\ k\Omega$ or $192\ k\Omega$	
$C_1 = 5\ nF$	
$C_2 = \dfrac{0.01}{4\pi \times 10^3 \times 1.59 \times 10^4} = 50\ pF$	

It can be seen here that the resistor values are very large, and although a MICROCAP or SPICE simulation may give acceptable results, in practice the circuit may fail to meet the specification.

We may effect a better design by changing the scaling factors. Choose $C_1 = 0.1\ \mu F$ then

$$k_m = \frac{1}{4\pi \times 10^3 \times 10^{-7}} = 796$$

and the new values are

$R_1 = 10.6\ k\Omega$ or $656\ \Omega$	
$R_2 = 105\ k\Omega$ or $6.56\ k\Omega$	
$R_3 = 597\ \Omega$ or $9.6\ k\Omega$	
$C_1 = 0.1\ \mu F$	
$C_2 = 1\ nF$	

Note the 100:1 capacitor spread and 10:1 spread of resistor values.

The circuit with the response is shown in Figure 4.22.

4.6.2 Gain and phase shift of the circuit

From equation (4.32) we may write

$$\left| \frac{V_o}{V_i}\ (j\omega) \right| = \frac{K b_0 \omega_0^2}{[(b_0 \omega_0^2 - \omega^2) + (b_1 \omega \omega_0)^2]^{0.5}}$$

$$\phi(j\omega) = 180° - \tan^{-1} \frac{b_1 \omega \omega_0}{b_0 \omega_0^2 - \omega^2}$$

By considering the three conditions $\omega = 0$, $\omega = \omega'$, $\omega = \infty$, a reasonable idea of the variation of gain and phase with variation of ω may be obtained and the response is shown in figure 4.23. Note that the MICROCAP simulation gives $\phi(j\omega)$ from $-180°$ to $-360°$; this is because of the formulation of its programs and is consistent with the result shown in figure 4.23.

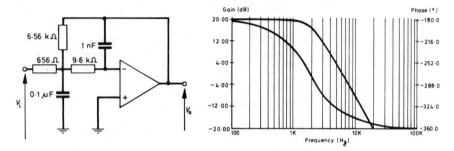

Figure 4.22

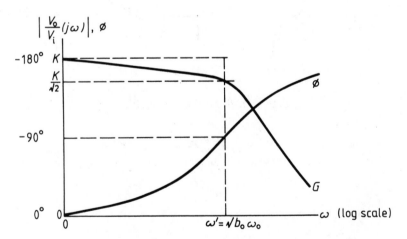

Figure 4.23 Gain/phase response for low-pass filter

$$\omega = 0 \quad \left| \frac{V_o}{V_i}(j0) \right| = K \text{ (the dc gain)}; \ \phi(j0) = 180°$$

$$\omega = \infty \quad \left| \frac{V_o}{V_i}(j\infty) \right| \rightarrow 0; \ \phi(j\infty) \rightarrow 0°$$

$$\omega = \omega' \quad \left| \frac{V_o}{V_i} (j\omega') \right| = K \Bigg/ \frac{b_0}{b_1} \; ; \; \phi(j\omega') = 90° \quad \text{where } \omega' = \sqrt{b_0} \, \omega_0$$

Referring to figure 4.21, it can be seen that at low frequencies capacitors C_1 and C_2 present high reactance values and the signal passes with very little attenuation, the gain being given by $H(j\omega) = -R_2/R_1$. At high frequencies the reactances of C_1 and C_2 are low, the signal being by-passed to ground via C_1. There are again two roll-off circuits to give the prescribed second-order response.

4.6.3 The multiple feedback high-pass filter

The configuration is shown in figure 4.24 from which also note that the resistor/capacitor positions are interchanged when compared with the low-pass configuration.

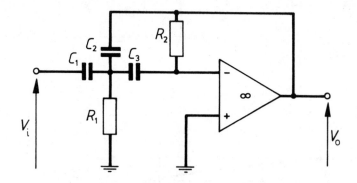

Figure 4.24   MFB high-pass filter

The admittances are: $Y_1 = j\omega C_1$, $Y_2 = j\omega C_3$, $Y_3 = j\omega C_2$, $Y_4 = 1/R_2$, $Y_5 = 1/R_1$. Insertion of these admittances into equation (4.30) yields

$$\frac{V_o}{V_i} (j\omega) = \frac{\omega^2 \dfrac{C_1}{C_2}}{-\omega^2 + j\omega \dfrac{(C_1 + C_2 + C_3)}{C_2 C_3 R_2} + \dfrac{1}{R_1 R_2 C_2 C_3}} \tag{4.38}$$

which may be compared with the standard form

$$\frac{V_o}{V_i} (j\omega) = \frac{K\omega^2}{-\omega^2 + jB_1 \omega\omega_0 + B_0 \omega_0^2} \tag{4.39}$$

From which we obtain the following relationships:

$$K = \frac{C_1}{C_2}$$

$$B_1 \omega_0 = \frac{(C_1 + C_2 + C_3)}{C_2 C_3 R_2}$$ (4.40)

$$B_0 \omega_0^2 = \frac{1}{R_1 R_2 C_2 C_3}$$

Consider normalised values of $C_1 = C_3$ and ω_0, then equations (4.40) reduce to

$$C_2 = \frac{1}{K}$$

$$R_2 = \frac{(2K + 1)}{B_1}$$

$$R_1 = \frac{K}{B_0 R_2}$$

To illustrate the design method, we will consider a simple example.

Example 4.6

Consider the previous specifications for the VCVS filter (example 4.4) which is summarised: $K = 10$, $A_{max} = 3$ dB, $A_{min} \geqslant 40$ dB, $f_c = 20$ kHz, $f_s = 2$ kHz.

The low-pass Chebyshev coefficients are: $b_1 = 0.65$, $b_0 = 0.708$, scaling factor 796.

Initially the coefficients b_0, b_1 must be converted into their equivalent high-pass values and this is achieved as follows:

$$B_1 = \frac{b_1}{b_0}, \quad B_0 = \frac{1}{b_1}$$

Therefore

$$B_1 = \frac{0.65}{0.708} = 0.918, \quad B_0 = \frac{1}{0.708} = 1.412$$

Select $C_1 = C_3 = 0.01 \ \mu F$

$$C_2 = \frac{0.01}{10} = 0.001 \ \mu F$$

$$R_2 = \frac{21}{0.918} = 22.88 \ \Omega$$

$$R_1 = \frac{10 \times 0.708}{22.88} = 0.309 \ \Omega \quad \text{normalised}$$

Applying the scaling factor to the normalised resistor values produces $R_1 = 246 \ \Omega$, $R_2 = 18.2$ kΩ. The circuit with its response is shown in figure 4.25.

An alternative approach is to convert the high-pass specification into the equivalent low-pass specification and then design the prototype low-pass. Finally, application of the $CR:RC$ transformation produces the prototype high-pass. It is recommended that the reader adopts this approach when designing high-pass filter circuits.

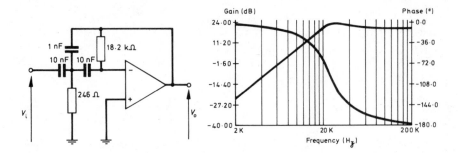

Figure 4.25   Circuit and response for MFB high-pass filter

Consider the specification 'translated' to the equivalent low-pass specification in figure 4.26.

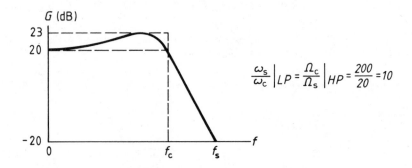

$$\frac{\omega_s}{\omega_c}\bigg|_{LP} = \frac{\Omega_c}{\Omega_s}\bigg|_{HP} = \frac{200}{20} = 10$$

Figure 4.26   Specification for example 4.6

Using the low-pass design with $b_1 = 0.65$, $b_0 = 0.708$:

$$C_2 \leqslant \frac{0.65^2}{4 \times 11 \times 0.708} < 0.0135 \text{ F}$$

Select $C_2 = 0.01$ F:

$$R_2 = \frac{\dfrac{0.65}{0.01} \pm \sqrt{\left[\left(\dfrac{0.65}{0.01}\right)^2 - \dfrac{4 \times 11 \times 0.708}{0.01}\right]}}{2 \times 0.708} = 69.4 \ \Omega \text{ or } 22.38 \ \Omega$$

giving $R_1 = 6.94 \ \Omega$ or $2.238 \ \Omega$.

Consideration of component magnitudes suggests we select $R_1 = 2.238\ \Omega$, $R_2 = 22.38\ \Omega$ and choose $C = 0.01\ \mu\text{F}$.

Scaling factors are

$$k_f = 2\pi \times 2 \times 10^4 = 4\pi \times 10^4$$

$$k_m = \frac{1}{4\pi \times 10^4 \times 10^{-8}} = 796$$

$$R_3 = \frac{1}{b_0 R_2 C_2} = \frac{1}{0.708 \times 22.38 \times 0.01} = 6.31\ \Omega$$

We now convert these components into the equivalent high-pass components.

Noting that $C_{\text{LP}} \to R_{\text{HP}} = \dfrac{1}{C_{\text{LP}}}$ and

$$R_{\text{LP}} \to C_{\text{HP}} = \frac{1}{R_{\text{LP}}}$$

$$R_1 = \frac{1}{1} = 1\ \Omega, \quad R_2 = \frac{1}{0.01} = 100\ \Omega$$

$$C_3 = \frac{1}{6.31}\ \text{F}$$

$$C_1 = \frac{1}{2.24}\ \text{F}, \quad C_2 = \frac{1}{22.4}\ \text{F}$$

The scaled values are

$$R_1 = 796\ \Omega, \quad R_2 = 79.6\ \text{k}\Omega, \quad C_3 = 1.58\ \text{nF}, \quad C_1 = 4.46\ \text{nF}, \quad C_2 = 446\ \text{pF}$$

and the final circuit with its response is shown in figure 4.27.

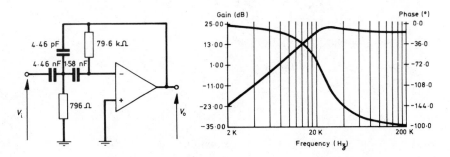

Figure 4.27   Circuit and responses for example 4.6

Note that the capacitor values are rather small and the spread between the two resistor values is 100:1 and between the capacitor values 3.5:1. The circuit operation is the inverse of the low-pass filter. At low frequencies, capacitors C_1 and C_3 block the signal and at high frequencies allow the signal through with negligible signal loss.

4.6.4 The multiple feedback band-pass filter

The concept of an infinite gain VCVS circuit is particularly useful when applied to the MFB band-pass circuit in figure 4.28.

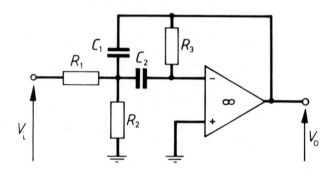

Figure 4.28   MFB band-pass filter

The conditions imposed on the admittances are: $Y_1 = 1/R_1$, $Y_2 = j\omega C_2$, $Y_3 = j\omega C_1$, $Y_4 = 1/R_3$, $Y_5 = 1/R_2$ which when inserted into equation (4.30) gives, after manipulation, the expression

$$\frac{V_o}{V_i}(j\omega) = \frac{-j\dfrac{\omega}{C_1 R_1}}{-\omega^2 + \dfrac{j\omega(C_1 + C_2)}{C_1 C_2 R_3} + \dfrac{(R_1 + R_2)}{C_1 C_2 R_1 R_2 R_3}} \tag{4.41}$$

The standard form is given by the expression

$$\frac{V_o}{V_i}(j\omega) = \frac{-\dfrac{j\omega\omega_0 G}{Q}}{-\omega^2 + \dfrac{j\omega\omega_0}{Q} + \omega_0^2} \tag{4.42}$$

A comparison of equation (4.41) and equation (4.42) produces for normalised C_1 and ω_0:

$$Q = \frac{R_3 C_2}{1 + C_2}$$

$$\frac{1}{R_1} + \frac{1}{R_2} = C_2 R_3 = Q(1 + C_2) \tag{4.43}$$

$$\frac{1}{R_1} = \frac{G}{Q}$$

Noting that $G_1 = \dfrac{1}{R_1}$ etc., subsequent manipulation of equation (4.43) gives

$$G_2 = Q(1 + C_2) - G_1 = Q(1 + C_2) - \frac{G}{Q}$$

For G_2 to be real and positive, then conditions must be imposed in the inequality

$$C_2 \geqslant \frac{G}{Q^2} - 1 \tag{4.44}$$

Example 4.7

A band-pass filter is required to have a centre frequency gain of 20 dB, a band-width of 500 Hz and a centre frequency of 1 kHz. The design is to use readily available capacitors.

$$G = \text{Antilog } \frac{20}{20} = 10$$

$$Q = \frac{f_0}{\Delta f} = \frac{1000}{500} = 2$$

Select $C_1 = 1$ F, giving $C_2 \geqslant \dfrac{10}{4} - 1 \geqslant 1.5$ F.

Choose $C_2 = 2$ F, then

$$G_2 = 2(3) - \frac{10}{2} = 1 \text{ s}$$

$$G_1 = \frac{G}{Q} = \frac{10}{2} = 5 \text{ s}$$

$$G_3 = \frac{C_2}{Q(1 + C_2)} = \frac{1}{3} \text{ s}$$

Scaling: $k_f = 2\pi \times 10^3$, $k_m = 1.59 \times 10^4$ for $C_1 = 0.01 \ \mu\text{F}$.

$$R_1 = 1.59 \times 10^4 \times \frac{1}{5} = 3.14 \text{ k}\Omega$$

$R_2 = 1.59 \times 10^4 \times 1 = 15.9\ \text{k}\Omega$

$R_3 = 1.59 \times 10^4 \times 3 = 47.7\ \text{k}\Omega$

$C_2 = 0.02\ \mu\text{F}$

$C_1 = 0.01\ \mu\text{F}$

The circuit with its response is shown in figure 4.29.

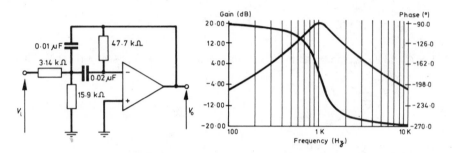

Figure 4.29   Circuit and responses for MFB band-pass filter

It is left as an exercise for the reader to show that the gain and phase response take the forms as for the VCVS band-pass circuit with the exception that the phase response extends from $-90°$ down to $-270°$. The filtering action is similar to the previously considered VCVS circuit but note that at low frequencies C_2 blocks the signal and at high frequencies C_1 is a short-circuit, producing a unity gain circuit.

The principal advantages of the MFB circuits over the VCVS is that they are less sensitive to component variations (see chapter 8) and have one component fewer within the design. A disadvantage is that they produce an additional phase shift of $180°$.

4.7 Higher-order filter circuits

It was mentioned earlier in chapter 3 that a principal advantage of a VCVS circuit was the good circuit isolation properties it gave. These properties are usefully employed when cascading sections to produce higher-order filters, the basic principle being shown in figure 4.30.

Here

$$H_1(j\omega) = |H_1(j\omega)|,\ \phi_1(j\omega)\ \text{etc.}\quad \text{from which}$$

$$\frac{V_{o_3}}{V_i} = H(j\omega) = H_1(j\omega) \times H_2(j\omega) \times H_3(j\omega)\quad \text{where}$$

$$|H(j\omega)| = |H_1| \times |H_2| \times |H_3| \tag{4.45}$$

$$\phi(j\omega) = \phi_1 + \phi_2 + \phi_3$$

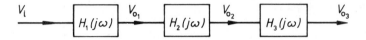

Figure 4.30 Cascaded filter sections

For example, a unity-gain third-order filter constructed from cascaded first-order and second-order circuits would provide $|H| = 1$, $\phi = -270°$.

If an *even*-order function is required, then there will be one or more even-order stages, whereas if the order is *odd*, one first-order stage in combination with one or more even-order stage(s) will be required. Although only active circuit realisations are to be considered, it should be noted that a combination of active and passive circuits may be employed. A third-order single amplifier filter circuit is shown in figure 4.31, which illustrates an alternative technique.

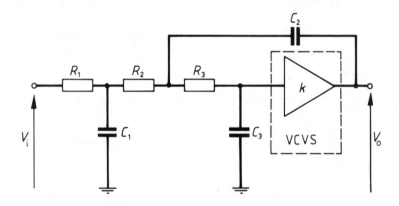

Figure 4.31 Third-order low-pass circuit

An advantage of the cascaded section technique is that each circuit (or active resonator) is tuned to the same frequency. A distinct disadvantage is their rather poor sensitivity response to small changes in circuit parameters and amplifier gain; this imposes a limit on the number of circuits which may be cascaded.

Example 4.8

A fourth-order, low-pass, Butterworth filter is required to have a gain of 16 and a cut-off frequency of 1 kHz. VCVS circuits are to be used with 0.01 μF capacitors.

We may represent the filter as two dissimilar sections as shown in figure 4.32. The operator s is used in order to correspond with tables in chapter 2, which is seen to be a fourth-order function whose roots may be obtained using the equations from chapter 2. We will use the results from that section which are summarised in table 2.1.

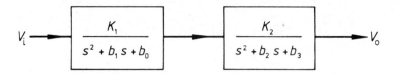

Figure 4.32 Block-diagram representation of example 4.7

Here

$$\frac{V_o}{V_i}(s) = \frac{K_1 K_2}{(s^2 + b_1 s + b_0)(s^2 + b_2 s + b_3)}$$

We see that the appropriate coefficients are $b_1 = 0.766$, $b_0 = 1$; $b_2 = 1.848$, $b_3 = 1$.

We will assign a gain of 4 to each stage and summarise the data as follows:

First stage	*Second stage*
$K_1 = 4, b_1 = 0.766, b_0 = 1$	$K_2 = 4, b_2 = 1.848, b_3 = 1$

The design procedure is identical with that covered in example 4.2.

First stage

$$R_2 = \frac{(0.766 \pm (0.766^2 - 4(-2))^{0.5})}{2} = 1.85 \ \Omega$$

$$R_1 = \frac{1}{1.85} = 0.541 \ \Omega$$

$$C_1 = C_2 = 1 \ F$$

$$R_3 = \frac{4}{3} \times 2.39 = 3.187 \ \Omega, \ R_4 = 3 \times 3.187 = 9.56 \ \Omega$$

Second stage

$$R_2 = \frac{(1.848 \pm (1.848^2 - 4(-2))^{0.5})}{2} = 2.613 \ \Omega$$

$$R_1 = \frac{1}{2.613} = 0.383 \ \Omega$$

$$R_3 = \frac{4}{3} \times 2.996 = 3.99\ \Omega$$

$$R_4 = 3 \times 3.99 = 11.97\ \Omega$$

$$C_1 = C_2 = 1\ F$$

Scaling:

$$k_f = 2\pi \times 10^3$$

$$k_m = \frac{10^8}{2\pi \times 10^3} = 1.59 \times 10^4$$

The final scaled circuit with its response is shown in figure 4.33. Note well from the design that the two sections which comprise the filter are not identical because of the coefficients (b_1, b_2) being unequal.

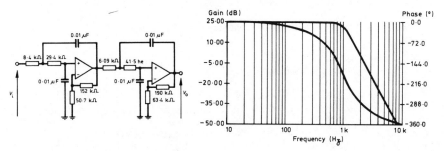

Figure 4.33   Circuit and gain/phase responses for example 4.8

Example 4.9

The specification for a high-pass filter is as follows:

(a) maximum loss in the pass-band is to be 3 dB down to 2 kHz;
(b) minimum loss at 250 kHz to be greater than 60 dB;
(c) the high-frequency gain is to be 20 dB.

The data suggests a Chebyshev response. From equation (2.15) the order n is obtained and note that $\Omega_c/\Omega_s = 8 = \omega_s/\omega_c$:

$$n = \frac{\cosh^{-1}\left[\dfrac{(10^6 - 1)}{(10^{0.3} - 1)}\right]^{0.5}}{\cosh^{-1} 8} = 2.74$$

Round up to 3.
    Using equation (2.14):

$$dB = 10\log_{10}\left[1 + \{\cosh 3\cosh^{-1} 8\}^2\right] = 66\ dB$$

The design data are now translated into the equivalent low-pass and the steps followed as shown in example 4.3.

We note that $\Omega_c/\Omega_s = 8 = \omega_s/\omega_c$, and for $\omega_c = 1$ rad/s normalised $\omega_s = 8$ (16 kHz).

We shall require a first-order section and a second-order section cascaded as shown in figure 4.34.

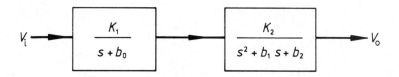

Figure 4.34   Cascaded low-pass equivalent for example 4.9

Combining these we obtain the transfer function for the filter:

$$\frac{V_o}{V_i} = \frac{K_1 K_2}{(s + b_0)(s^2 + b_1 s + b_2)} \tag{4.46}$$

From table 2.4 the appropriate coefficients for $A_{max} = 3$ dB are obtained as $b_0 = 0.298, b_1 = 0.298, b_2 = 0.839$. The parameters for the two stages are

First stage   *Second stage*
$K_1 = 2, b_0 = 0.298$ $K_2 = 5, b_1 = 0.298, b_2 = 0.839$

First stage
This employs the VCVS circuit of example 4.1.

$$R_1 = \frac{1}{b_0} = \frac{1}{0.298} = 3.36 \ \Omega$$

$$C = 1 \ F$$

$$R_2 = R_3 = KR_1 = 6.72 \ \Omega$$

Second stage
This uses the circuit of example 4.2.

$$R_2 = \frac{(0.298 \pm (0.298^2 + 4(0.839 \times 3))^{0.5})}{2 \times 0.839} = 2.07 \ \Omega$$

$$R_1 = \frac{1}{2.07 \times 0.839} = 0.576 \ \Omega$$

$$R_3 = \frac{5}{4} \times 2.646 = 3.31 \ \Omega$$

$R_4 = 5 \times 2.646 = 13.23\ \Omega$

$C_1 = C_2 = 1\ F$

Use the $RC:CR$ transformation outlined in example 4.3 to obtain the low-pass to high-pass transformation of components as follows:

First stage	*Second stage*
$R_1 = 1\ \Omega$	$R_1 = R_2 = 1\ \Omega$

$$C = \frac{1}{3.36} = 0.298\ \Omega \qquad\qquad C_1 = \frac{1}{0.576} = 1.736\ F$$

$$R_2 = R_3 = 6.72\ \Omega \qquad\qquad C_2 = \frac{1}{2.07} = 0.483\ F$$

$$R_3 = 3.31\ \Omega$$

$$R_4 = 13.23\ \Omega$$

Scaling:

$$k_f = 4\pi \times 10^3$$

$$\text{Choose } C = \frac{10}{2 \times 10^3} = 5\ nF$$

$$k_m = \frac{5 \times 10^9}{4\pi \times 10^3} = 1.59 \times 10^4$$

The resistor values are multiplied by 1.59×10^4 and the capacitor values divided by 2×10^8.

The final circuit with its response is shown in figure 4.35.

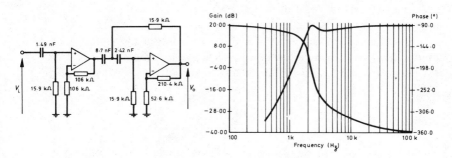

Figure 4.35   Circuit and responses for third-order high-pass filter

We noted earlier that cascaded filter sections possess poor sensitivities to component variations. Another design problem results in difficulties in the tuning of these circuits because of the large number of components used to realise the particular design.

4.8 Conclusions

To conclude this chapter, we note that a number of popular circuits have been studied using the positive gain VCVS and negative gain (infinite gain) MFB concepts. We noted the merits of the VCVS in that they have an easier closed-loop gain adjustment, a higher input impedance and a low output impedance, which makes for easier cascading of sections without the need for isolation transformers. The MFB circuit has one fewer component, perhaps two in the band-pass configuration, and is also easy to tune. Because of the lower input impedance, this circuit may suffer when used in cascade to produce a higher-order filter.

Later, in chapter 8, it will be seen that the VCVS filter has poor Q and ω_0 sensitivities with gain, which imposes restrictions on the Q-values that may be realised.

A brief mention should be made here concerning the variation of input impedance with frequency throughout the frequency range of the filter. Since we are using reactive elements and an op-amp whose gain varies with frequency, it should come as no surprise to find that the input impedance may swing from being inductive reactive to capacitive reactive for a wide variation in the input signal frequency.

A final point to consider concerns the accuracy of the design when variation in the filter parameters, caused by a finite and perhaps low value of A_0, is taken into account. We saw in chapter 2 that for so long as A_0 is large then the behaviour of the circuit is independent of the characteristics of the amplifier and dependent only on the external circuit components. When A_0 becomes small (less than 500 for a 741), then the situation changes and the circuit becomes more dependent on the characteristics of the amplifier.

Any design and calculations must therefore make an allowance for this condition. Briefly stated, in the low-pass VCVS design for example, the gain constant K will be modified, as also will the values of the coefficients b_1 and b_0.

Problems

4.1. Design a low-pass filter to have the response characteristic shown in figure 4.36. Use the design equations to confirm (or otherwise) the pass-band and stop-band attenuations. Design both VCVS and MFB filters using 0.1 μF capacitors.
[*Ans.* VCVS: 2.93 kΩ, 38.4 kΩ, 0.1 μF, 414 kΩ, 45.6 kΩ.
MFB: 65.6 kΩ, 656 kΩ, 8.49 kΩ, 0.1 μF, 2 nF]

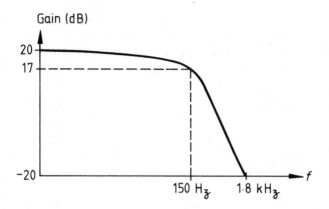

Figure 4.36

4.2. A filter is to have a gain of 5 and a permitted ripple width of 1 dB, the end of the ripple occurring at 2.4 kHz. The gain is to be at least −26 dB down at 200 Hz. Design the prototype low-pass and high-pass circuits and obtain the scaled circuit using 0.01 μF capacitors. Check out the pass-band and stop-band attenuations for the order of the filter chosen and sketch the expected frequency response. Design both VCVS and MFB circuits.
   [*Ans.* VCVS: 22.9 nF, 4.26 nF, 6.7 kΩ, 6.7 kΩ, 21.7 kΩ, 87 kΩ.
   MFB: 10 nF, 2 nF, 1.1 nF, 6.67 kΩ, 333 kΩ]

4.3. A filter is required to have a gain of 18 dB with a permitted pass-band ripple of 3 dB up to 6.8 kHz. The output is to be −80 dB down at 1 kHz. Design the filter detailing *all* the stages in the design including the confirmation of the actual pass-band and stop-band attenuations. Use VCVS circuits with gains of 2 and 4 respectively, using 2 nF capacitors throughout.
   [*Ans.* 1st stage: 340 pF, 10.6 nF, 11.7 kΩ, 11.7 kΩ, 142 kΩ, 142 kΩ.
   2nd stage: 1.72 nF, 450 pF, 11.7 kΩ, 11.7 kΩ, 87 kΩ, 261 kΩ]

4.4. A high-pass filter is to be designed to be maximally flat Butterworth having a gain of 20 dB in the pass band, a cut-off frequency of 5 kHz and a roll-off of 100 dB/decade in the transition band. Design the filter using a low-pass prototype employing 2 nF capacitors. Use VCVS circuits throughout and make use of the *RC*:*CR* transformation.
   [*Ans.* 1st stage: 15.9 kΩ, 2 nF, 31.8 kΩ, 31.8 kΩ.
   2nd stage: 15.9 kΩ, 15.9 kΩ, 1.78 nF, 2.25 nF, 57.9 kΩ, 71.6 kΩ.
   3rd stage: 15.9 kΩ, 15.9 kΩ, 3.5 nF, 1.14 nF, 66.9 kΩ, 82.7 kΩ]

4.5. A band-pass filter is required having a centre frequency gain of 20 dB, a centre frequency of 500 Hz and a pass band of 100 Hz. Design the VCVS circuit using $K = 2$, and capacitors selected from $C = 10/f_0$ μF.
   [*Ans.* 15.9 kΩ, 10.09 kΩ, 41 kΩ, 82 kΩ, 82 kΩ, 0.02 μF, 0.02 μF]

4.6. Design a band-pass filter circuit to have a centre frequency of 1 kHz, a centre frequency gain of 20 dB with a Q-factor of 2. Use an MFB circuit and select one of the capacitors to be 0.1 μF.

[*Ans.* 314 Ω, 1.59 kΩ, 4.7 kΩ, 0.1 μF, 0.2 μF]

5 Biquadratic Filter Circuits

5.1 Introduction

In this chapter, we will consider a number of interesting circuits associated with biquadratic functions. Initially, the R–C type will be discussed and the analysis conducted using an approach resembling that used for analogue computing circuitry. It should be pointed out here that some authors prefer to use state-space methods of analysis for these circuits, and for this the reader is referred to the relevant literature.

Next we consider an increasingly popular type of filter based on the fabrication of op-amps and capacitors on the same chip, the capacitors being switched to simulate resistor values. Such *switched capacitor* circuits are becoming increasingly popular in communications, spectrum analysers and medical electronics where a degree of programmability may be achieved.

Finally, brief consideration is given to the concept of an *all-resistor* biquad which uses the roll-off of the op-amp at high frequencies which was considered in chapter 3.

5.2 The biquadratic circuit

This concept provides for stable and easily tuned filter circuits. They require more components than the MFB and VCVS circuits previously considered, but in its bandpass form it is capable of producing much greater Q-factors than the VCVS and MFB circuits. Furthermore, with the use of the *quad* op-amp chip a *universal biquad* circuit is realisable which means that the five basic filter forms may be obtained using appropriately selected component values. The name *biquad* is coined because the general circuit can be shown to be defined (mathematically) as the ratio of two quadratic expressions.

A starting point for the analysis will use the transfer function for the high-pass filter:

$$H(j\omega) = \frac{K\omega^2}{-\omega^2 + \left(\dfrac{\omega_0}{Q}\right) j\omega + \omega_0^2} \tag{5.1}$$

and if we write $s = j\omega$ this becomes

$$H(s) = \frac{-Ks^2}{s^2 + \left(\dfrac{\omega_0}{Q}\right) s + \omega_0^2}$$

and to simplify the manipulation we will normalise the angular frequency to $\omega_0 = 1$ rad/s. Therefore

$$H(s) = \frac{-Ks^2}{s^2 + \dfrac{s}{Q} + 1} \tag{5.2}$$

This expression may be re-written as

$$H(s) = \frac{V_o}{V_i}(s) = \frac{-K}{1 + \dfrac{1}{Qs} + \dfrac{1}{s^2}} \times \frac{\left(\dfrac{P}{K}\right)}{\left(\dfrac{P}{K}\right)}$$

where, note that the expression, although modified, is mathematically unaltered and which may be considered to be equivalent to

$$V_o = -P$$

$$V_i = \frac{P}{K} + \frac{P}{K}\left(\frac{1}{Qs}\right) + \frac{P}{K}\left(\frac{1}{s^2}\right) \tag{5.3}$$

or

$$P = KV_i - \frac{P}{Qs} - \frac{P}{s^2} = KV_i + \frac{V_o}{s} + \frac{V_o}{s^2}$$

This equation states that an output $P = -V_o$ is the sum of three input voltages and this specification may be realised by the circuit shown in figure 5.1.

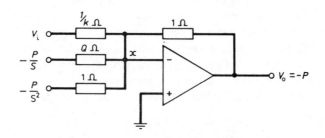

Figure 5.1   Simple summing amplifier

Nodal analysis at x yields, assuming $V_x = 0$:

$$(0 - V_\mathrm{i})K + \left(0 + \frac{P}{s}\right)\frac{1}{Q} + \left(0 + \frac{P}{s^2}\right) + (0 - V_\mathrm{o}) = 0$$

or

$$-KV_\mathrm{i} = V_\mathrm{o} + \frac{V_\mathrm{o}}{Qs} + \frac{V_\mathrm{o}}{s^2} = V_\mathrm{o}\left(1 + \frac{1}{Qs} + \frac{1}{s^2}\right)$$

$$\frac{V_\mathrm{o}}{V_\mathrm{i}} = \frac{-Ks^2}{s^2 + \dfrac{s}{Q} + 1}$$

From equation (5.3) it will be seen that we require to produce voltages $(-P/s = V_\mathrm{o}/s)$ and $(-P/s^2 = V_\mathrm{o}/s^2)$, and it should be noted that $(1/s)$ implies an integrator which has both R and C set to unit value of 1 ohm and 1 Farad. Similarly, $(1/s^2 = 1/s \times 1/s)$ implies a cascading of two identical integrator circuits which combine to produce the second-order effect.

It was mentioned earlier that the circuit synthesis would be similar to the method applied to the analysis of analogue computing circuits and with this in mind a convenient starting point will be a consideration of the circuit shown in figure 5.2.

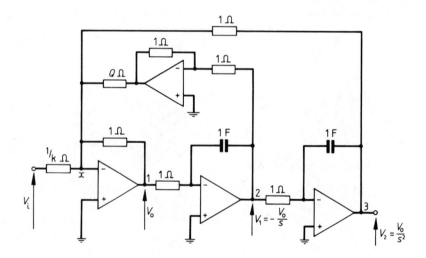

Figure 5.2 Circuit realisation for the basic biquad

Obviously, the transfer function $H(s)$ demands that we produce a second-order function which we see is satisfied by cascading two identical integrator sections. A further requirement is the summation of the three quantities given by equation (5.3), which are mixed at node x to the input of the first op-amp. Note further that this circuit, and indeed most biquad circuits, employ op-amps in the inverting mode and therefore an *odd* number of stages must be used to avoid the possibility of positive feedback arising with subsequent instability occurring. Note also that the inner-loop op-amp also produces negative feedback, the signal then being applied to node x through resistor Q ohms.

Nodal equation at x yields

$$(0 - V_i)K + (0 - V_o) + \left(0 - \frac{V_o}{s}\right)\frac{1}{Q} + \left(0 - \frac{V_o}{s^2}\right) = 0$$

or

$$-KV_i = V_o \left(1 + \frac{1}{Qs} + \frac{1}{s^2}\right)$$

and

$$\frac{V_o}{V_i} = \frac{-Ks^2}{s^2 + \dfrac{s}{Q} + 1}$$

The circuit shown in figure 5.2 will, therefore, represent the high-pass function (output at node 1).

Further, noting that

$$V_1 = -\frac{V_o}{s}, \quad V_2 = \frac{V_o}{s^2} = -\frac{V_1}{s}$$

at node 2

$$\frac{V_o}{V_i} \times \frac{V_1}{V_o} = \frac{V_1}{V_i} = \frac{-Ks^2}{s^2 + \dfrac{s}{Q} + 1} \times \left(\frac{-1}{s}\right) = \frac{Ks}{s^2 + \dfrac{s}{Q} + 1}$$

which is a band-pass function. Also

$$\frac{V_2}{V_i} = \frac{V_1}{V_i} \times \frac{V_2}{V_1} = \frac{Ks}{s^2 + \dfrac{s}{Q} + 1} \times \left(\frac{-1}{s}\right)$$

$$= -\frac{K}{s^2 + \dfrac{s}{Q} + 1}$$

which is a low-pass function.

If only low-pass and band-pass configurations are required, one of the op-amps may be eliminated by combining the operations of summation and integration as shown in figure 5.3.

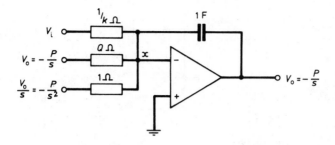

Figure 5.3 Summation and integration

At node x, nodal analysis yields

$$(0 - V_i)K + (0 - V_o)\frac{1}{Q} + \left(0 - \frac{V_o}{s}\right) + (0 - V_o)s = 0$$

or

$$-KV_i = V_o \left(s + \frac{1}{s} + \frac{1}{Q}\right)$$

giving

$$\frac{V_o}{V_i} = \frac{-Ks}{s^2 + \dfrac{s}{Q} + 1} \tag{5.4}$$

It can be seen that three inputs must be provided $V_i; -\dfrac{P}{s} = V_o; -\dfrac{P}{s^2} = \dfrac{V_o}{s}$ and clearly to obtain V_o/s a further integrator will be required followed by an inverter circuit to give the correct polarity. The final circuit is shown in figure 5.4 which also shows the appropriate voltages and their polarities.

$$\frac{V_1}{V_i} = \frac{-Ks}{s^2 + \dfrac{s}{Q} + 1} \qquad \text{band-pass}$$

$$\frac{V_2}{V_i} = \frac{V_1}{V_i} \times \frac{V_2}{V_1} = \frac{-Ks}{s^2 + \dfrac{s}{Q} + 1} \times \left(-\frac{1}{s}\right) = \frac{K}{s^2 + \dfrac{s}{Q} + 1} \qquad \text{low-pass}$$

We will now consider in detail two useful circuits employing the biquad concept.

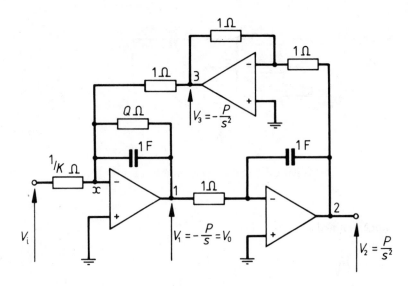

Figure 5.4   Circuit realisation of low-pass and band-pass cases

5.3 The band-pass circuit

If we are only interested in the band-pass circuit, figure 5.4 may be re-drawn as shown in figure 5.5, which includes the modification shown in figure 5.3. Note that one of the separate integrators has been removed (since we are combining summation and integration at the input) and that this particular configuration has the advantage of providing a positive-going output.

It is left as an exercise for the reader to show that consideration of conditions at node x and $\dfrac{V_o}{V_i} = \dfrac{V_y}{V_i} \times \dfrac{V_o}{V_y}$ produces the relationship

$$\frac{V_o}{V_i} = \frac{Ks}{s^2 + \dfrac{s}{Q} + 1} \tag{5.5}$$

Consider now the circuit with actual component values and consider further the conditions $C_1 = C_2 = C$ and $R_4 = R_5 = R_6$. From figure 5.5, nodal analysis at x yields

$$\frac{V_i}{R_1} + \left(sC + \frac{1}{R_2}\right)V_y + \frac{V_\omega}{R_3} = 0$$

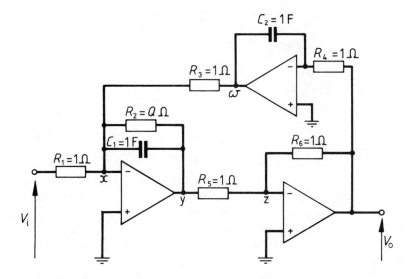

Figure 5.5   Biquad band-pass circuit

also

$$V_z = -V_y, \; V_\omega = - \frac{V_z}{sCR_4} = \frac{V_y}{sCR_4}$$

which finally gives

$$\frac{V_o}{V_i} = \frac{\dfrac{s}{CR_1}}{s^2 + \dfrac{s}{CR_2} + \dfrac{1}{C^2 R_3 R_4}} \qquad (5.6)$$

Writing $s = j\omega$, equation (5.6) may be re-written in the standard form

$$\frac{V_o}{V_i}(j\omega) = \frac{j\omega G \dfrac{\omega_0}{Q}}{-\omega^2 + j\omega \dfrac{\omega_0}{Q} + \omega_0^2} \qquad (5.7)$$

where

$$\omega_0 = \frac{1}{(C^2 R_3 R_4)^{0.5}}$$

$$G = \frac{R_2}{R_1}$$

$$R_2 = \frac{Q}{C\omega_0}$$ (5.8)

$$R_1 = \frac{Q}{CG\omega_0}$$

$$R_3 = R_4 = \frac{1}{\omega_0 C}$$

for R_3 made equal to R_4.

The filter may be tuned by adjusting R_3 to fix the centre frequency, adjustment of R_2 fixes the Q-factor, and finally the centre frequency gain G is established by adjusting R_1. The tuning condition for the circuit is often referred to as *orthogonal tuning* as distinct from *iterative tuning* where continuous adjustment is made until all the specification requirements are met. Orthogonal tuning is always preferable to iterative tuning on the grounds of convenience and the saving of time.

We will now consider an example to illustrate the design technique.

Example 5.1

A band-pass filter is required having a centre frequency of 1500 Hz and a bandwidth of 300 Hz, the centre frequency gain being 5. Obtain the scaled circuit design.

Initially, normalise the centre frequency and capacitor values, that is $\omega_0 = 2\pi f_0 = 1$ rad/s, $C = 1$ F. Equations (5.8) reduce to

$$R_3 = R_4 = \frac{1}{\omega_0} = 1, \ R_3 = \frac{1}{R_4}, \ R_1 = \frac{Q}{G}, \ R_2 = Q$$

From chapter 4, equation (4.26):

$$Q = \frac{f_0}{\Delta f} = \frac{1500}{300} = 5$$

$$G = 5$$

$$R_1 = \frac{5}{5} = 1 \ \Omega$$

$$R_2 = 5 \ \Omega$$

$$R_3 = R_4 = 1 \ \Omega$$

We must now choose C, and we use $C = (10/f_c) \ \mu F = 10/1500 = 0.0067 \ \mu F$ which is difficult to achieve, so we choose $C = 0.01 \ \mu F$ from which the scaling factors are

$$k_f = \frac{\omega_0}{\omega_n} = \frac{2\pi \times 1500}{1} = 3\pi \times 10^3$$

$$k_m = \frac{1}{3\pi \times 10^3 \times 10^{-8}} = \frac{10^5}{3\pi} = 1.06 \times 10^4$$

The scaled circuit values become:

$C = 0.01\ \mu F$

$R_1 = R_3 = R_4 = 10.6\ k\Omega$

$R_2 = 53\ k\Omega$

The complete circuit with its response is shown in figure 5.6.

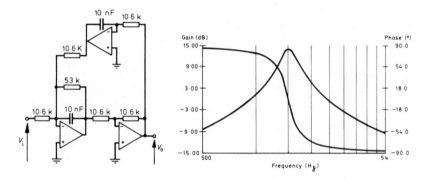

Figure 5.6 Band-pass biquad

It will be observed from the transfer function given by equation (5.7) that the phase shift will vary between $+90°$ ($\omega = 0$) through $0°$ ($\omega = 3000\pi$) to $-90°$ ($\omega \to \infty$).

5.4 The band-reject, notch or band-elimination circuit

It was mentioned earlier that with the introduction of the four op-amp (quad) chip, all of the five basic filters could be realised within the framework of a so-called *universal* filter. Consideration is now given to the band-reject circuit and figure 5.4 is re-drawn with the inclusion of an additional summing amplifier, as shown in figure 5.7.

The band-reject circuit, of which there are several forms (see figures 5.22 and 5.23), is particularly useful in applications where a specific frequency is to be removed. An example would be the removal of the hum associated with the 50 Hz power frequency in electronic medical equipment or instrumentation systems. Such a band-reject filter, as shown in figure 5.7, uses the output from the previously considered band-pass circuit, which is added to the input signal via the summing amplifier which then inverts the combined signal.

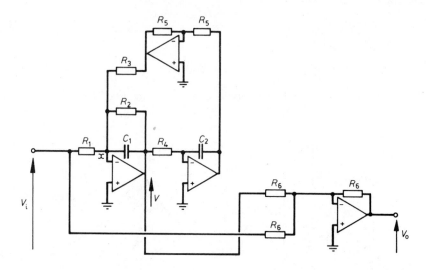

Figure 5.7   Band-reject filter circuit

It is left as an exercise for the reader to show that

$$\frac{V_1}{V_i}(s) = \frac{-\dfrac{s}{C_1 R_1}}{s^2 + \dfrac{s}{C_1 R_2} + \dfrac{1}{C_1 C_2 R_3 R_4}}$$

(5.9)

and

$$V_o = -(V_i + V_1)$$

or

$$\frac{V_o}{V_i}(s) = -\left(1 + \frac{V_1}{V_i}\right) = -\left(1 - \frac{\dfrac{s}{C_1 R_1}}{s^2 + \dfrac{s}{C_1 R_2} + \dfrac{1}{C_1 C_2 R_3 R_4}}\right)$$

$$\frac{V_o}{V_i}(s) = -\left(\frac{s^2 + \dfrac{s}{C_1 R_2} - \dfrac{s}{C_1 R_1} + \dfrac{1}{C_1 C_2 R_3 R_4}}{s^2 + \dfrac{s}{C_1 R_2} + \dfrac{1}{C_1 C_2 R_3 R_4}}\right) \qquad (5.10)$$

If we impose the condition $R_1 = R_2$ ($G = 1$) then the expression reduces to

$$\frac{V_o}{V_i}(s) = \frac{-(s^2 + \omega_0^2)}{s^2 + \dfrac{s}{C_1 R_2} + \omega_0^2} \qquad (5.11)$$

where

$$\omega_0^2 = \frac{1}{C_1 C_2 R_3 R_4}$$

Further, write $s = j\omega$ and noting that the denominator has the form

$$s^2 + s\left(\frac{\omega_0}{Q}\right) + \omega_0^2$$

it can be seen, comparing this with equation (5.11)

$$\frac{\omega_0}{Q} = \frac{1}{C_1 R_2}$$

producing

$$Q = \frac{C_1 R_2}{(C_1 C_2 R_3 R_4)^{0.5}} = \left(\frac{R_2^2 C_1}{C_2 R_3 R_4}\right)^{0.5}$$

Normally $C_1 = C_2 = C$ and the expressions become

$$\omega_0 = \frac{1}{C(R_3 R_4)^{0.5}}, \quad Q = \frac{R_2}{(R_3 R_4)^{0.5}}$$

Furthermore, if we make $R_3 = R_4$ then

$$\omega_0 = \frac{1}{CR_3}, \quad Q = \frac{R_2}{R_3}$$

and by fixing C, ω_0 is adjusted by R_3 and Q by R_2. Now

$$\left|\frac{V_o}{V_i}(j\omega)\right| = \frac{\omega_0^2 - \omega^2}{\left\{(\omega_0^2 - \omega^2)^2 + \left(\dfrac{\omega}{C_1 R_2}\right)^2\right\}^{0.5}}$$

and it can be seen that at selected frequencies

$$\left|\frac{V_o}{V_i}(j0)\right| = 1, \quad \left|\frac{V_o}{V_i}(j\infty)\right| = 1, \quad \left|\frac{V_o}{V_i}(j\omega_0)\right| = 0 \quad \text{or}$$

$$0 \text{ dB}, \qquad\qquad 0 \text{ dB}, \qquad -\infty \text{ dB}$$

The suggestion here is that for the condition $\omega = \omega_0$, the circuit will provide infinite attenuation. In practice, this is not the case because of the imperfections within the components. It is interesting to consider the phase produced by the circuit at the selected frequencies:

$$\phi(j\omega) = \phi_1 - \phi_2 \quad \text{where} \quad \phi_1 = \angle\omega_0^2 - \omega^2$$

$$\phi_2 = \tan^{-1}\left(\frac{\dfrac{\omega}{C_1 R_2}}{\omega_0^2 - \omega^2}\right)$$

It is useful to draw these as independent angles and then to combine them as shown in figure 5.8, which also shows the gain response.

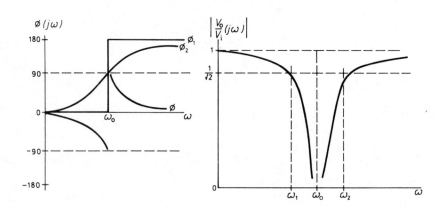

Figure 5.8   Gain/phase responses of the biquad band-reject

At the 3 dB points, bandwidth = $\dfrac{\omega_0}{Q}$ = $\omega_2 - \omega_1$.

Example 5.2

An undesirable 50 Hz signal is required to be removed from the output of an electrocardiogram. The filter must pass frequencies above 55 Hz and below 45 Hz with no more than 3 dB of loss. The low- and high-frequency loss must be 0 dB.

From the specifications we see, using equation (4.26) from chapter 4, that

$$Q = \frac{50}{(55 - 45)} = 5 = \frac{R_2}{R_3}$$

$$\omega_0 = 2\pi \times 50 = \frac{1}{CR_3}$$

Normalise $\omega_0 = 1$ rad/s, $C = 1$ F, and using the previously derived equations:

$$R_3 = R_4 = 1 \; \Omega$$

$$\frac{R_2}{R_1} = G$$

and

$$G = 1$$

Therefore $R_2 = Q \; \Omega = R_1$.

Let $R_5 = R_6 = 1 \; \Omega$ $R_3 = R_4 = 1 \; \Omega$ $R_1 = R_2 = Q \; \Omega$ $C_1 = C_2 = 1$ F.

Scaling: $k_f = \dfrac{2\pi \times 50}{1} = 100\,\pi$

Choose $C = 1 \; \mu$F: $k_m = \dfrac{1}{100\pi \times 10^{-6}} = 3180$

Then $R_3 = R_4 = R_5 = R_6 = 3.18$ kΩ
     $R_1 = R_2 = 15.9$ kΩ
     $C_1 = C_2 = 1 \; \mu$F.

The circuit with its response is shown in figure 5.9.

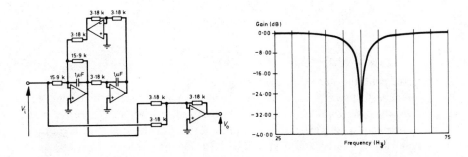

Figure 5.9   Circuit and response for example 5.2

A further note concerning the band-reject circuit

Re-call equation (5.11) which may be re-written

$$\frac{V_o}{V_i}(s) = \frac{-(s^2 + \omega_p^2)}{s^2 + \left(\dfrac{\omega_0}{Q}\right)s + \omega_0^2} \tag{5.12}$$

where ω_p is the frequency at which the maximum attenuation occurs. It should be noted that $\omega_0 < \omega_p < \omega_0$ and three principal conditions arise:

$\dfrac{\omega_p}{\omega_0} < 1$ which defines a *high-pass* notch

$\dfrac{\omega_p}{\omega_0} = 1$ which defines a *regular* notch

$\dfrac{\omega_p}{\omega_0} > 1$ which defines a *low-pass* notch

If we consider the magnitude characteristic of equation (5.12):

$$\left|\frac{V_o}{V_i}(j\omega)\right| = \frac{\omega_p^2 - \omega^2}{\left\{(\omega_0^2 - \omega^2)^2 + \omega^2\left(\dfrac{\omega_0}{Q}\right)^2\right\}^{0.5}}$$

It can be seen that three principal points arise:

$$\left.\left|\frac{V_o}{V_i}(j\omega_p)\right|\right|_{\omega=\omega_p} = 0$$

$$\left.\left|\frac{V_o}{V_i}(j0)\right|\right|_{\omega=0} = \left(\frac{\omega_p}{\omega_0}\right)^2$$

$$\left.\left|\frac{V_o}{V_i}(j\infty)\right|\right|_{\omega\to\infty} = 1$$

The three conditions are illustrated in figure 5.10.

Referring to figure 5.10, which shows the disposition of the three conditions given by the ratio ω_p/ω_0, it can be seen that only for the condition $\omega_p = \omega_0$ does symmetry of the magnitude response occur. For the remaining conditions, asymmetrical responses are produced with the low-pass notch exhibiting gain up to the rejection frequency range and the high-pass notch exhibiting attenuation. The reader is directed to example **5.6**, figure 5.23, for an illustration of the three notch conditions.

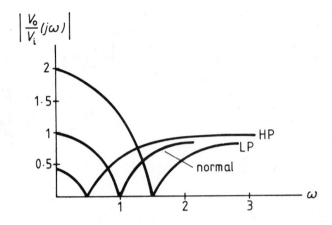

Figure 5.10   Normal, high- and low-pass notch responses

5.5 Switched capacitor filters

These are sometimes referred to as *analogue sampled data filters* and use MOS technology. Using current IC technology it is quite difficult to achieve accurate values of resistance and capacitance which exhibit good linearity and stability with variation of temperature. Most of the active R–C filters which have been completely integrated require careful trimming and consequently are too expensive for commercial mass production.

One approach which makes active filters more sympathetic towards IC technology is the active R circuit, which although it dispenses with the need for capacitors, still depends on resistor tolerances and op-amp gain–bandwidth products. A big disadvantage is that resistors require more chip area than is the case with capacitors.

The true switched capacitor filter in many ways overcomes most of the aforementioned problems. In MOS technology the switched capacitor can be implemented using two MOSFETs which are operated using a two-phase clock system, the clock phases not overlapping (see figure 5.11).

It is possible to achieve high precision in realising capacitor ratio values and, moreover, since MOS capacitors approach closely ideal capacitor characteristics, greater linearity and stability are possible than with diffused resistors. It is therefore possible to achieve fully integrated and stable active filter circuits using the switched capacitor principle.

The basic principle of switched capacitor action may be understood by considering the two-phase, non-overlapping MOSFET clock system shown in figure 5.11.

With ϕ_1 high, C is connected to V_i. When ϕ_2 is high, C is connected to V_o, otherwise C is connected to neither V_i nor V_o. The principle of operation is now discussed with reference to figure 5.12.

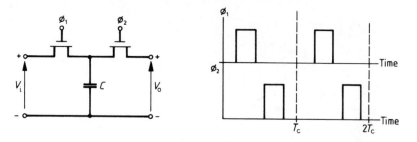

Figure 5.11   MOSFET two-phase clock

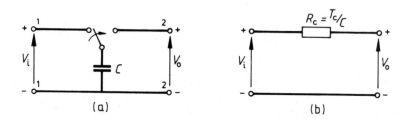

Figure 5.12   Switched capacitor with circuit equivalent

The single-pole double-throw (SPDT) switch is assumed to be driven using ϕ_1, ϕ_2 in figure 5.11 and it will be assumed that the switch is initially connected to position 1. We will also assume that V_i is a function of time and that the periodic time T of V_i is much greater than the switching frequency T_c of the SPDT switch.

In position 1, capacitor C charges to V_i and with the switch made to position 2, the capacitor discharges (or charges) to V_o.

The charge transferred in T_c seconds is given by

$$q(t) = C(V_i - V_o)$$

and

$$i(t) = \frac{\Delta q}{\Delta t} = \frac{C}{T_c}(V_i - V_o) \tag{5.13}$$

A resistor required to give the same magnitude of current will have a value given by the expression

$$R_c = \frac{V_i - V_o}{i} = \frac{T_c}{C(V_i - V_o)}$$

where

$$R_c = \frac{T_c}{C} = \frac{1}{f_c C} \qquad\qquad (5.14)$$

and

$$f_c = \frac{1}{T_c}$$

It may be concluded that the circuit shown in figure 5.12(a) performs the same function as a resistor having a value T_c/C ohms, as shown in figure 5.12(b). If the switching frequency is much larger than the signal frequencies being handled, then the equivalence of switched capacitor with resistor is accurate enough. Should the frequencies be of the same order, however, then data-sampling techniques must be employed.

A simple circuit with which to illustrate the procedure is the inverting integrating amplifier shown in figure 5.13.

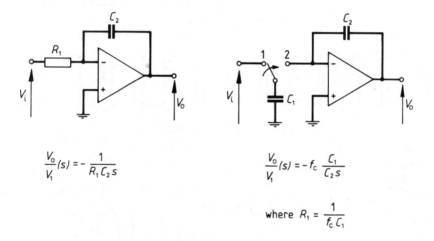

Figure 5.13 Integrator circuit with switched capacitor equivalent

With the switch in position 1, C_1 charges to V_i and when the switch is moved to position 2, this charge is removed from C_1 and transferred to C_2. It should be noted that all of the charge is transferred because the virtual-earth condition of the op-amp constrains the voltage across C_1 to be zero. The resultant charge on C_2 is the charge previous to making the switch *less* the charge transferred from C_1. This charge is subtracted because of the inverting operation of the amplifier.

It can be seen that the time constant of the integrator may easily be adjusted by means of the clock frequency.

The nature of the operation of a sampled data integrator circuit may be further illustrated by reference to figure 5.14.

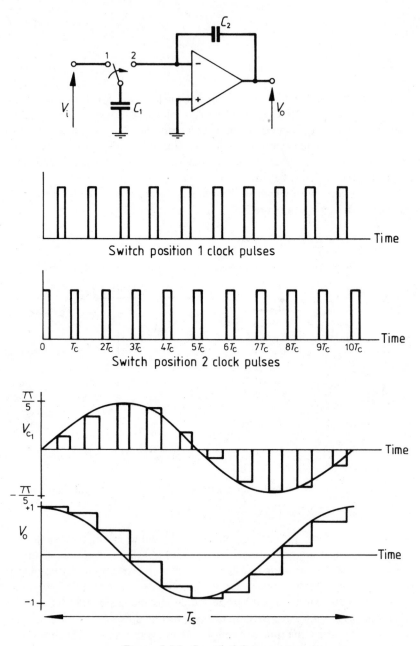

Figure 5.14 Sampled data concept

The input is assumed to be sinusoidal having a frequency f_s and a peak amplitude of 1 V. The input is sampled in the centre of each clock cycle and this value (V_i) is held on the capacitor C_1 for half a clock period and then is transferred to C_2 when the switch is made to position 2. The output is held on C_2 until a new sample of V_i is transferred to C_2 and so on. The clock frequency is made ten times larger than the signal frequency ($f_c = 10 \, f_s$); also, using the previous expression with $s = j\omega$ in order to consider circuit operation with frequency:

$$\frac{V_o}{V_i}(j\omega) = -f_c \frac{C_1}{j\omega C_2}$$

$$\left| \frac{V_o}{V_i}(j\omega) \right| = \frac{C_1}{C_2} \times \frac{f_c}{2\pi f_s}$$

when

$$\left| \frac{V_o}{V_i}(j\omega) \right| = 1$$

then

$$\frac{C_1}{C_2} = \frac{2\pi f_s}{f_c} = \frac{\pi}{5}$$

It can be shown that at time $t = nT_c$, the voltage transferred to C_2 is given by

$$V_o(nT_c) = V_o[(n-1)T_c] - \frac{C_1}{C_2} V_i[(n-1)T_c] \tag{5.15}$$

or voltage on C_2 = initial voltage − transferred voltage from C_1. The negative sign arises because of the inversion across the amplifier.

Assuming that at time $t = 0$, $V_o = 1$ V, the input is initially sampled at $t = T_c/2$ and is held on C_1 for a further period of time $t = T_c/2$ before the voltage on C_1 is transferred to C_2 at $t = T_c$.

From figure 5.14 it can be seen that

$$T_c = \frac{T_s}{10} = \frac{360°}{10} \quad \text{or} \quad \frac{T_c}{2} \equiv 18°$$

Using equation (5.15) we may compute the voltage V_o for any switching time, for example

$$t = \frac{T_c}{2} \equiv 18°$$

The sampled voltage on C_1 is $(\pi/5)\sin 18°$ and the voltage on C_2 for $T_c \leqslant t \leqslant 2T_c$ is given by

$$V_0(2T_c) = V_0(t = T_c) - \frac{\pi}{5} V_i \left(t = \frac{T_c}{2}\right)$$

$$= 1 - \frac{\pi}{5} \sin 18° = 0.806 \text{ V}$$

This voltage is held on C_2 until $t = 2T_c$ while the input voltage is being sampled at the new time $t = (3T_c)/2$ giving a new voltage on C_1 of $(\pi/5)\sin 54°$. The voltage on C_2 for $2T_c \leqslant t \leqslant 3T_c$ is given by

$$V_0(3T_c) = V_0(2T_c) - \frac{\pi}{5} V_i(2T_c)$$

$$= 0.806 - 0.508 = 0.298 \text{ V}$$

This process may be repeated for all the clock intervals to obtain the waveform shown in figure 5.14.

Consider now the circuits shown in figure 5.15, the second circuit being the switched capacitor equivalent of the first circuit.

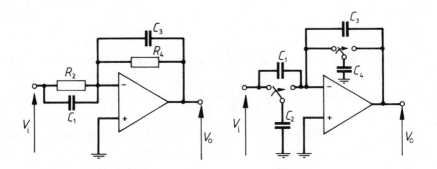

Figure 5.15   Further circuit with its switched capacitor equivalent

Here

$$C_2 = \frac{1}{f_c R_2}, \quad C_4 = \frac{1}{f_c R_4}$$

also

$$\frac{V_0}{V_i}(s) = -\frac{C_1}{C_3} \frac{\left(s + \dfrac{1}{R_2 C_1}\right)}{\left(s + \dfrac{1}{R_4 C_3}\right)}$$

and on substituting for C_2, C_4:

$$\frac{V_o}{V_i}(s) = -\frac{C_1}{C_3} \frac{\left(s + f_c \dfrac{C_2}{C_1}\right)}{\left(s + f_c \dfrac{C_4}{C_3}\right)} \tag{5.16}$$

To illustrate only, consider now an example of the technique outlined above.

Example 5.3

We wish to implement a switched capacitor design to give the specification shown in figure 5.16.

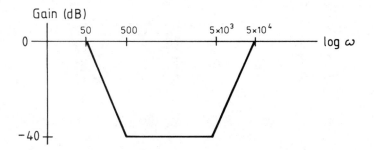

Figure 5.16   Band-reject specification

The transfer function for this band-reject response takes the form

$$\frac{V_o}{V_i}(s) = \frac{(s + 500)(s + 5 \times 10^3)}{(s + 50)(s + 5 \times 10^4)}$$

which may be broken down into the form

$$\frac{V_o}{V_i}(s) = \frac{(s + 500)}{(s + 50)} \times \frac{(s + 5 \times 10^3)}{(s + 5 \times 10^4)} \tag{5.17}$$

Select $C_1 = C_3$ for convenience and also assign a value of 10 pF to these components. Further, select the clock frequency f_c to be 10 kHz.

Equating the coefficients of equations (5.16) and (5.17) gives for the first section:

$$\frac{C_2}{C_1} = 0.05 \quad \text{or} \quad C_2 = 0.5 \text{ pF}$$

$$\frac{C_4}{C_3} = 0.005 \quad \text{or} \quad C_4 = 0.05 \text{ pF}$$

Repeating for the second section gives the capacitor values

$$C_2 = 5 \text{ pF}, \ C_4 = 50 \text{ pF}$$

It will be observed that there is a large *spread* of capacitor values and that the capacitor values used are relatively small. The final circuit is shown in figure 5.17.

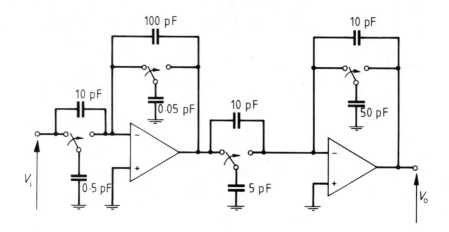

Figure 5.17   Switched capacitor realisation of band-reject filter

Conclusions

It is perhaps timely here to remind ourselves of some of the problems associated with integrated circuit switched-capacitor filters, of which the following are typical:

1. the MOS switch used in the implementation should have high off-resistance and low on-resistance, as well as having low parasitic capacitance values from the source and drain to the substrate;
2. should the clocking rate be of the order of signal frequency, then sampled data techniques must be used for the analysis, which means that the input signal should be band limited;
3. consideration of the circuits shows that the switched-capacitor resistor does not provide a continuous path for op-amp leakage currents, which are essential to minimise offsets at the input to the op-amp. This may complicate the design of circuits that require a precise value of closed-loop gain, such as is often required in VCVS filter circuits.

On the plus side we find that many modern communication systems can readily take advantage of switched-capacitor filters because the filter parameters of gain, Q-factor, centre frequency and bandwidth, are accurately controlled by capacitor ratios and clock frequency.

Typical devices manufactured currently are dual-in-line, that is, two separate filters which may be used singly or cascaded to produce all the five filter functions; the programming requirements are the selection of a clock frequency, power supplies and three or four resistors.

5.6 All-resistor active biquad circuits

It has been previously mentioned that at high frequencies the op-amp derived circuits cease to have benefits over the normal passive circuit design. The main cause of the op-amp limitation is due to the finite gain–bandwidth product imposed on the device which tends to move the location of the poles in the filter, which for certain Q-factors could well result in circuit instability. Also, as the signal frequency is increased, slew-rate limitation may occur because of the inability of the op-amp accurately to output the signal.

Consideration must be given to the use of inductors at these considerably higher working frequencies. Of course, inductors are a much more viable consideration at very high frequencies and are indeed extensively used. Nevertheless, integrated circuit technology has the advantage of compatibility between devices, and inductor technology, although advancing, is still not yet fully compatible with silicon chip technology. A starting point will be the one-pole roll-off model of the op-amp discussed in chapter 3.

$$A(j\omega) = \frac{A_0 \omega_a}{j\omega + \omega_a} = \frac{A_0}{1 + \dfrac{j\omega}{\omega_a}} \tag{5.18}$$

It may be observed that the gain/frequency relationship has a similar relationship to that of an elementary R-C network. The conclusion may therefore be drawn that biquad filters may be designed using only operational amplifiers and resistors with the exclusion of capacitors. A number of circuits have been designed by many workers around the model of the op-amp specified in equation (5.18) to work at relatively high frequencies in the range 20–300 kHz.

The circuit in figure 5.18 shows an active-R biquad circuit.

The constraints on R_4 and R_5 are given by the expressions

$$R_4 = (1 - \beta)R$$

$$R_5 = \beta R \tag{5.19}$$

$$\text{for } 0 < \beta < 1$$

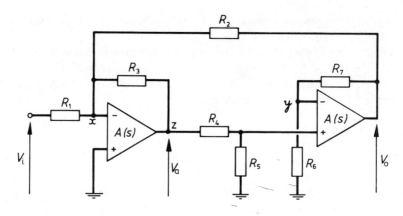

Figure 5.18   Active-R biquad circuit

Applying nodal analysis at points x, y and z within the circuit in figure 5.18 yields

$$\left.\begin{array}{l}
\dfrac{(V_i - V_x)}{R_1} + \dfrac{(V_z - V_x)}{R_2} + \dfrac{(V_z - V_x)}{R_3} = 0 \\[3mm]
\dfrac{(V_o - V_y)}{R_7} = \dfrac{V_y}{R_6} \\[3mm]
V_z = -A(s)V_x = -\dfrac{A_0\omega_a}{s + \omega_a}\, V_x \\[3mm]
V_o = \dfrac{A_0\omega_a}{s + \omega_a}\,(\beta V_a - V_y)
\end{array}\right\}
\qquad (5.20)$$

After considerable manipulation of equation (5.20), and by making the valid assumption that the open-loop gain (A_0) is much larger than the ratio of the resistor values, we finally obtain the desired transfer function:

$$\frac{V_z}{V_i}(s) = \frac{a_1 s + a_0}{s^2 + b_1 s + b_0} \qquad (5.21)$$

where

$$a_1 = \frac{A_0 \omega_a}{1 + \dfrac{R_1}{R_2} + \dfrac{R_1}{R_3}}$$

$$a_0 = \frac{(A_0 \omega_a)^2}{\left(1 + \dfrac{R_1}{R_2} + \dfrac{R_1}{R_3}\right)\left(1 + \dfrac{R_7}{R_6}\right)}$$

$$b_1 = A_0 \omega_a \left(\frac{1}{1 + \dfrac{R_7}{R_6}} - \frac{1}{1 + \dfrac{R_3}{R_2} + \dfrac{R_3}{R_1}}\right)$$

$$b_0 = \frac{\dfrac{R_3}{R_2} (\beta A_0^2 \omega_a^2)}{1 + \dfrac{R_3}{R_2} + \dfrac{R_3}{R_1}}$$

(5.22)

After re-arranging equations (5.22) and introducing a coefficient y, where $y > 1$ to ensure positive resistor ratios, we obtain the expressions

$$1 + \frac{R_1}{R_2} + \frac{R_1}{R_3} = \frac{A_0 \omega_a}{a_1}$$

$$1 + \frac{R_3}{R_2} + \frac{R_3}{R_1} = \frac{A_0 \omega_a R_3}{a_1 R_1}$$

$$\frac{R_7}{R_6} = \frac{A_0 \omega_a}{b_1 \left(1 - \dfrac{1}{y}\right)}$$

$$R_4 = R(1 - \beta)$$

$$R_5 = R\beta$$

$$\beta = \frac{b_0}{A_0 \omega_a \left(A_0 \omega_a - \dfrac{b_1}{y} - a_1\right)}$$

$$\frac{R_1}{R_2} = \frac{b_0}{A_0 \omega_a \beta a_1}$$

(5.23)

Consider now the design of an all op-amp band-pass filter.

Example 5.4

An all-resistor band-pass biquad circuit is to be designed to have the following specification: $Q = 5$, $G = 10$, $f_0 = 80$ kHz. Obtain the realisation of the filter assuming that the op-amp employed has the following parameters: $A_0 = 8 \times 10^4$, $\omega_a = 10\pi$ rad/s. The second pole of the op-amp is taken to be at $\omega = 8\pi \times 10^5$ rad/s.

Using the equation for a band-pass response:

$$\frac{V_o}{V_i}(j\omega) = \frac{j\omega\omega_0 \dfrac{G}{Q}}{-\omega^2 + \dfrac{j\omega\omega_0}{Q} + \omega_0^2}$$

Comparing with equation (5.21) with $a_0 = 0$, $s = j\omega$, the equation becomes, after insertion of the given data

$$\frac{V_o}{V_i}(j\omega) = \frac{j10^6\,\omega}{-\omega^2 + j10^5\,\omega + 2.53 \times 10^{11}}$$

where $a_1 = 10^6$, $b_1 = 10^5$, $b_0 = 2.53 \times 10^{11}$. Selecting $R = 2$ kΩ, equations (5.23) produce

$\beta = 0.07, R_1 = 145\ \Omega$ (selected), $R_2 = 100\ \Omega$

$R_3 = 2.07$ kΩ, $R_4 = 1.86$ kΩ, $R_5 = 140\ \Omega$, $R_6 = 120\ \Omega$ (selected)

$R_7 = 33$ kΩ

The final circuit with its response is shown in figure 5.19.

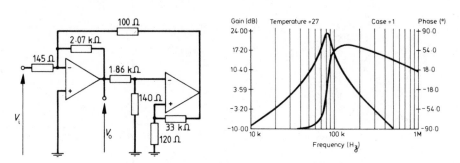

Figure 5.19   All-resistor band-reject circuit

5.7 Conclusions

It can readily be seen from the foregoing analysis that active-*R* circuits are more difficult to design and implement than the usual *R–C* biquad circuits previously considered. One might ask whether such circuits are really desirable, when at frequencies in excess of 100 kHz inductors become far more attractive components to use in terms of size, cost and ease of manufacture.

It was mentioned earlier that until inductors become more integrated, circuit-compatible active realisations will still be sought after. Nevertheless, active-*R* filters are almost impossible to produce on a production-line basis, mainly because of the sensitivities associated with the gain–bandwidth product ($A_0 \omega_a$) being very dependent on power supply and temperature variations. Also, since perfect matching of op-amps of the same type is virtually impossible, individual circuit tuning must always be applied. Slew-rate limiting of the op-amp also causes problems at high frequencies where even low signals could cause slewing problems.

Some workers in this field have used CCVS (current-controlled-voltage-source) and CCCS (current-controlled-current-source) op-amp circuits instead of the more normal VCVS circuit. These circuits have the benefit of low input impedance levels which can lead to filters having a much higher frequency range.

Problems

5.1. Using the circuit shown in figure 5.5, obtain the design of a band-pass filter having the following specification: centre frequency gain 10 dB, centre frequency 10 kHz, with a bandwidth of 250 Hz. Use 10 nF capacitors in the design.
[*Ans.* $R_1 = 2$ kΩ, $R_2 = 6.4$ kΩ, $R_3 = R_4 = 1.6$ kΩ]

5.2. Using the circuit shown in figure 5.7, design a band-reject filter having the following specification: centre frequency 1 kHz, three dB bandwidth 100 Hz, low- and high-frequency loss 0 dB. Use 0.1 μF capacitors in the design.
[*Ans.* $R_1 = R_2 = 15.9$ kΩ, $R_3 = R_4 = R_5 = R_6 = 1.59$ kΩ]

5.3. The circuit shown in figure 5.20 is that of a band-pass filter having adjustable *ganged* capacitors and a variable resistor R_1. Obtain the transfer function of the filter in the form:

$$\frac{V_o}{V_i}(j\omega) = \frac{j2\omega\omega_0/Q}{-\omega^2 + j\omega \dfrac{\omega_0}{Q} + \omega_0^2}$$

where $\omega_0^2 = \dfrac{1}{C^2 R^2}$, $Q = K$.

Design a filter to have a centre frequency of 10 kHz and a bandwidth of 1 kHz. Use 0.1 μF capacitors in the design.
[*Ans.* $R_1 = 1.59$ kΩ, $R = 159$ Ω, $C = 0.1$ μF]

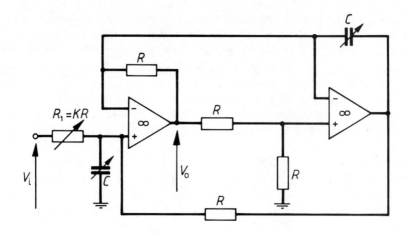

Figure 5.20

5.4. Obtain the transfer function $\dfrac{V_o}{V_i}$ $(j\omega)$ for the circuit shown in figure 5.21 in the form:

$$\frac{V_o}{V_i}\,(j\omega) = \frac{\omega^2 + j\,\dfrac{\omega}{C_2 R_2}\left(\dfrac{R_2}{R_3} - 1\right) - \dfrac{1}{C_1 C_2 R_1 R_2}}{-\omega^2 + j\,\dfrac{\omega}{C_2 R_2} + \dfrac{1}{C_1 C_2 R_1 R_2}}$$

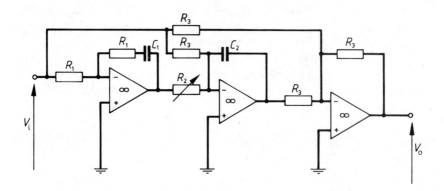

Figure 5.21

Examine the conditions when: (a) $R_2 = R_3$, (b) $R_2 = 2R_3$. Which responses do these conditions represent?

5.5. The circuit shown in figure 5.22 is that of a twin-T band-reject filter circuit employing two unity-gain op-amps. Write the nodal equations at the nodes 2, 3 and 4 and obtain the standard transfer function for the circuit, show-that $\omega_0 = \dfrac{1}{CR}$, $Q = \dfrac{1}{4(1-K)}$. Design a band-reject filter to have $Q = 10$, $f_0 = 1$ kHz and using 0.1 μF capacitors.
[*Ans. R* = 1.59 kΩ]

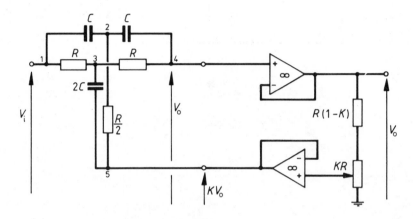

Figure 5.22

5.6. Obtain the transfer function for the circuit shown in figure 5.23 in the form:

$$\frac{V_o}{V_i}(j\omega) = \frac{\omega_z^2 - \omega^2}{-\omega^2 + j\omega\,\dfrac{\omega_0}{Q} + \omega_0^2}$$

where

$$\omega_z^2 = \frac{R_2}{R_1 R_3 R_5 C_1 C_2}, \quad \omega_0^2 = \frac{1}{R_4 R_5 C_1 C_2}$$

$$Q = \frac{R_5 R_6}{R_5 + R_6}\left(\frac{C_2}{R_4 R_5 C_1}\right)^{\frac{1}{2}}$$

Which filter response does the transfer function represent?

If the values of ω_0, R_1, C_1 and C_2 are normalised and the conditions $R_5 = R_6$, $R_2 = KR_1$ imposed, obtain the component values in terms of K, Q and ω_z. Explain the function of each of the variable components in the *tuning* of the circuit.

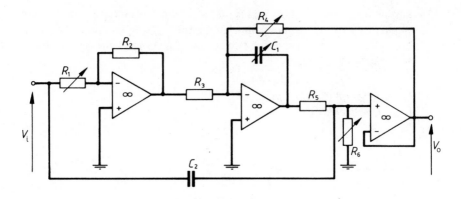

Figure 5.23

6   Passive Filter Circuit Design

6.1 Introduction

It would be misleading to suggest, with the advent of operational amplifiers and digital synthesis techniques, that passive filters have no role to perform in modern circuit applications.

At high frequencies they are usually the only practicable filter available, albeit in the form of coils, coaxial cables, waveguides or even in integrated circuit form.

The principal intention of this chapter is to show how the filter designer may utilise design data, usually in the form of tables, as an aid to designing a particular filter circuit. The data are usually in the form of *normalised* values of circuit capacitance and inductance appropriate to the order n and particular response requirement $H(j\omega)$ of the filter. Once the prototype low-pass circuit has been obtained, magnitude and frequency scaling is then applied to realise the final circuit. Another, important, reason for introducing passive circuit realisations is that from these circuits the gyrator-based equivalents considered in chapter 7 are derived.

Finally, a method of frequency transformation from low-pass to high-pass and band-pass is outlined which shows how prototype low-pass design data may be used to implement other responses.

6.2 The coefficient matching technique

This is a simple and effective means of obtaining a circuit transfer function, but is limited to simple structures. The method will be outlined by considering the circuit shown in figure 6.1. The circuit is of the second order, has two energy storage elements, and is terminated at both ends in resistors R_s (source) and R_L (load).

Analysis of the circuit yields:

$$\frac{V_o}{V_i}(s) = \frac{1}{R_s + sL + \dfrac{R_L}{1 + sCR_L}}\left(\frac{\dfrac{1}{Cs}}{R_L + \dfrac{1}{Cs}}\right)R_L$$

$$\frac{V_o}{V_i}(s) = \frac{\dfrac{1}{LC}}{s^2 + s\,\dfrac{(L + CR_LR_s)}{LCR_L} + \dfrac{(R_L + R_s)}{LCR_L}} \qquad (6.1)$$

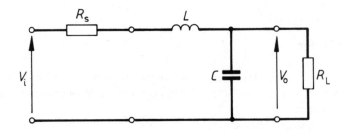

Figure 6.1 L–C section with resistor terminations

If now we assume the specifications of a second-order Butterworth response, then

$$\frac{V_o}{V_i}(s) = \frac{K}{s^2 + \sqrt{2}s + 1} \qquad (6.2)$$

Treating these expressions as being identical, we may therefore compare coefficients. Furthermore, it is usual to work with normalised terminating resistors $(R_s = R_L = 1\ \Omega)$:

$$\frac{L + C}{LC} = \sqrt{2}, \quad \frac{2}{LC} = 1, \quad K = \frac{1}{LC}$$

From which we obtain $L = \sqrt{2}$ H, $C = \sqrt{2}$ F.

The method appears elegant and straightforward, but becomes increasingly difficult to apply as the order n of the circuit increases.

There now follows the outline of a technique whic'· may be used to obtain the normalised values of inductance and capacitance for a ցiven network and a given response. The main aim here is to demonstrate how tables of design data are established and through the use of these tables passive filter realisations may be achieved.

Readers who require a deeper and more detailed account of the technique are referred to the appropriate texts listed in the Bibliography.

6.3 The design of doubly terminated L–C filters

The method is based on the extensive use of modern network theory and was originally outlined by Darlington and other workers. The design strategy involves the integration of the following concepts: insertion loss, reflection coefficient, network input impedance and the appropriate response requirement of the filter. Consider the loss-less network with resistive terminations shown in figure 6.2.

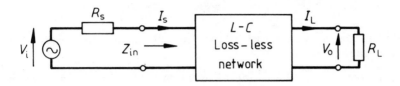

Figure 6.2 Loss-less network, doubly terminated

Any *L-C* network may be characterised at the source end of the network by a reflection coefficient ρ expressed as:

$$\rho(s) = \frac{Z_{IN}(s) - R_s}{Z_{IN}(s) + R_s} \tag{6.3}$$

Assuming a loss-less network and unequal terminations, then the power in the load may be expressed by:

$$P_L = \left| \frac{V_i R_L}{R_s + R_L} \right|^2 \bigg/ R_L \tag{6.4}$$

Maximum power occurs when the terminating resistors are equal ($R_L = R_s$):

$$P_{max} = \frac{V_o^2}{R_L} = \left(\frac{V_i}{2} \right)^2 \bigg/ R_s \tag{6.5}$$

Using these equations we obtain:

$$\frac{P_L}{P_{max}} = \frac{4 R_L R_s}{(R_s + R_L)^2} \tag{6.6}$$

Using definitions of insertion loss, reflection coefficient and response, it can be shown for an input impedance specified as

$$Z_{IN}(j\omega) = R + jX$$

associated with a loss-less circuit working under maximum power conditions ($R_L = R_s$):

$$\rho(s)\,\rho(-s) = 1 - \frac{4 R_L R_s}{(R_L + R_s)^2} \, |H(s)|^2 \tag{6.7}$$

where $H(s)$ is the appropriate circuit response.

Furthermore, by making $R_L = R_s = 1 \ \Omega$, equations (6.3) and (6.7) yield

$$\rho(s)\,\rho(-s) = 1 - |H(s)|^2 = \left| \frac{Z_{IN}(s) - 1}{Z_{IN}(s) + 1} \right|^2$$

from which is finally obtained:

$$Z_{IN}(s) = \frac{1 + \rho(s)}{1 - \rho(s)} \qquad (6.8)$$

We may now detail the basic strategy for obtaining the *L-C* ladder circuits from the given circuit requirements:

1. obtain the relevant $|H(s)|^2$ for the circuit, that is Butterworth or Chebyshev;
2. obtain $\rho(s)\rho(-s) = 1 - |H(s)|^2$;
3. deduce $Z_{IN}(s)$;
4. expand $Z_{IN}(s)$ into a continued fraction and obtain the ladder configuration of the normalised circuit;
5. scale the circuit to obtain the specified realisation.

To illustrate the use of the technique outlined above, we will now consider a number of examples.

Example 6.1

A third-order Butterworth filter is required normalised at 1 rad/s and terminated in equal 1 Ω resistors. Obtain the normalised low-pass design.

Using equation (2.2) from chapter 2 with $\epsilon = 1, s = j\omega, n = 3$:

$$|H(s)|^2 = \frac{1}{1 + (-js)^6} = \frac{1}{1 - s^6}$$

$$\rho(s)\,\rho(-s) = 1 - \frac{1}{1 - s^6} = \frac{s^6}{s^6 - 1}$$

$$= \left(\frac{s^3}{s^3 - 1} \right) \left(\frac{s^3}{s^3 + 1} \right)$$

from which

$$\rho(s) = \frac{s^3}{(s - P_1)(s - P_2)(s - P_3)} = \frac{s^3}{s^3 + 2s^2 + 2s + 1} \quad \text{(see table 2.1)}$$

where P_1, P_2, P_3 are the poles of the denominator, the denominator of this expression being obtained from table 2.1.

Inserting the expression for $\rho(s)$ into equation (6.8) gives, after rationalisation:

$$Z_{IN}(s) = \frac{2s^3 + 2s^2 + 2s + 1}{2s^2 + 2s + 1}$$

Finally, the continued division process is applied as follows:

$$\begin{array}{r} s \\ 2s^2 + 2s + 1 \overline{\smash{\big)}\ 2s^3 + 2s^2 + 2s + 1} \end{array}$$

$$2s^3 + 2s^2 + s$$

$$\begin{array}{r} 2s \\ s + 1 \overline{\smash{\big)}\ 2s^2 + 2s + 1} \end{array}$$

$$2s^2 + 2s$$

$$\begin{array}{r} s \\ 1 \overline{\smash{\big)}\ s + 1} \end{array}$$

$$\frac{s}{1}$$

The two circuits which will satisfy the requirement are shown in figure 6.3.

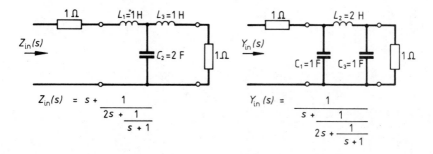

$$Z_{in}(s) = s + \cfrac{1}{2s + \cfrac{1}{s + 1}}$$

$$Y_{in}(s) = \cfrac{1}{s + \cfrac{1}{2s + \cfrac{1}{s + 1}}}$$

Figure 6.3 Prototype T and π circuits

By continuing to increase the order n of the response, a table of component values may be drawn up as shown in table 6.1. Of course, the procedure outlined above may be applied to deduce any loss-less configuration, but this is unnecessary once the table of component values has been drawn up. All that will be required is to scale the normalised values appropriate to the quoted filter specifications.

The method is also valid for unequal terminations and the reader is directed towards the more detailed literature on the topic.

Finally, by substituting

$$|H(s)|^2 = \frac{1}{1 + \epsilon^2 C_n^2(s)}$$

for the Chebyshev response, we may use the technique to determine an appropriate table for the inductors and capacitors for the Chebyshev circuit. It should be noted that there will be several such tables corresponding to the particular ripple value ϵ chosen, or more usually, the value of A_{MAX} permitted in the pass band.

Table 6.1 Element values for Butterworth filters normalised for 1 radian/s,
1 ohm, 1 ohm

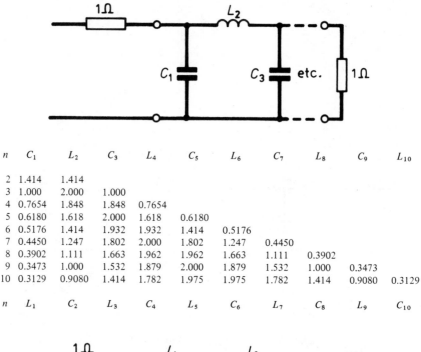

n	C_1	L_2	C_3	L_4	C_5	L_6	C_7	L_8	C_9	L_{10}
2	1.414	1.414								
3	1.000	2.000	1.000							
4	0.7654	1.848	1.848	0.7654						
5	0.6180	1.618	2.000	1.618	0.6180					
6	0.5176	1.414	1.932	1.932	1.414	0.5176				
7	0.4450	1.247	1.802	2.000	1.802	1.247	0.4450			
8	0.3902	1.111	1.663	1.962	1.962	1.663	1.111	0.3902		
9	0.3473	1.000	1.532	1.879	2.000	1.879	1.532	1.000	0.3473	
10	0.3129	0.9080	1.414	1.782	1.975	1.975	1.782	1.414	0.9080	0.3129
n	L_1	C_2	L_3	C_4	L_5	C_6	L_7	C_8	L_9	C_{10}

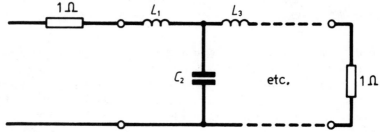

Example 6.2

Using the data in table 6.1 obtain the prototype low-pass circuit for a sixth-order filter. Scale the circuit to have a cut-off frequency of 100 Hz and to have equal 600 Ω terminations.

We will use the π-circuit; it is left as an exercise for the reader to take the T-circuit equivalent. This example will serve to illustrate the use of tables of data in designing passive, ladder-type filters.

Referring to table 6.1 we may obtain the normalised values of capacitance and inductance. For example, using the π-circuit equivalent the first component is a capacitor $C_1 = 0.5176$ F followed by a series inductor $L_2 = 1.414$ H which is then followed by a shunt capacitor $C_3 = 1.932$ F and so on. The values are listed below and are shown in figure 6.4:

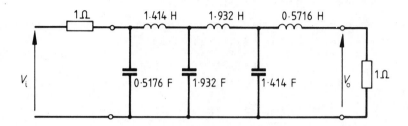

Figure 6.4 Prototype sixth-order low-pass

$C_1 = 0.5176$ F, $L_2 = 1.414$ H, $C_3 = 1.932$ F, $L_4 = 1.932$ H, $C_5 = 1.414$ F, $L_6 = 0.5176$ H

Scaling:

$$k_f = \frac{2\pi \times 100}{1} = 200\pi \quad L_o = L_n \left(\frac{k_m}{k_f}\right)$$

$$k_m = 600 \quad C_o = \frac{C_n}{k_m k_f}$$

The final scaled circuit is shown in figure 6.5.

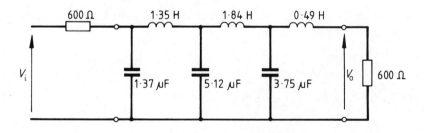

Figure 6.5 Scaled sixth-order circuit

An example will now be given to show how the method may be used to design passive Chebyshev filter circuits.

Example 6.3

A filter is to be designed around the following specifications:

Pass band up to 5 kHz with a permitted ripple of 3 dB.
Stop band should have a minimum loss greater than 100 dB at 30 kHz. The
terminations are to be 600 Ω pure resistors.

See chapter 2 for formulae used and it is recommended that the reader make
a brief review of example 2.4 before proceeding further:

$$A_{max} = 10\log_{10}(1 + \epsilon^2) = 3$$

from which $\epsilon = 1$:

$$n \geqslant \cosh^{-1} \left[\frac{(10^{A\,min/10} - 1)}{(10^{A\,max/10} - 1)}\right]^{\frac{1}{2}} \bigg/ \cosh^{-1} \frac{\omega_s}{\omega_p}$$

$$\geqslant \cosh^{-1} 10^5 / \cosh^{-1} 6 = 4.92$$

We take $n = 5$:

$$C_5^2(s) = -(16s^5 + 20s^3 + 5s)^2 \text{ (see table 2.2 where } s = j\omega)$$

$$|H(s)|^2 = \frac{1}{1 + \epsilon^2 C_5^2(s)}$$

$$\rho(s)\rho(-s) = 1 - |H(s)|^2 = \frac{\epsilon^2 C_5^2(s)}{1 + \epsilon^2 C_5^2(s)}$$

which yields on substituting $C_5^2(s)$:

$$\rho(s)\rho(-s) = \frac{(16s^5 + 20s^3 + 5s)^2}{(16s^5 + 20s^3 + 5s)^2 - 1} = \frac{N(s)}{D(s)}$$

The zeros of $D(s)$ and the zeros of $N(s)$ have now to be found. Since n is odd
the angles may be found using:

$$\theta_k = \left(\frac{2k + 1}{2n}\right) \pi; \ k = 0, 1, 2$$

and there will be a pole at the origin:

$$\theta_0 = \frac{\pi}{5}, \ \theta_1 = \frac{3\pi}{10}, \ \theta_2 = \frac{\pi}{2}$$

There will be poles of $\rho(s)\rho(-s)$ in the form $(s - P_0)(s - P_1)(s - P_2)$, obtained
as follows:

$$a = \frac{1}{n} \sinh^{-1}\left(\frac{1}{\epsilon}\right) = 0.177$$

$P_0 = -\sinh 0.177 \sin 18° \pm j\cosh 0.177 \cos 18°$

$\quad = -0.0547 \pm j0.9652$

$P_1 = -\sinh 0.177 \sin 54° \pm j\cosh 0.177 \cos 54°$

$\quad = -0.143 \pm j0.597$

$P_2 = -\sinh 0.177 \sin 90° \pm j\cosh 0.177 \cos 90°$

$\quad = -0.177$

Therefore

$$D(s) = (s + 0.177)\,(s^2 + 0.286s + 0.375)\,(s^2 + 0.1094s + 0.966)$$

$$= s^5 + 0.572s^4 + 1.442s^3 + 0.559s^2 + 0.419s + 0.064$$

The zeros of $N(s)$ may be found by noting that $s = j\omega$ and that

$$N(j\omega) = |\, C_5^2(\omega)|^2$$

The zeros occur for the condition $(16\omega^5 - 20\omega^3 + 5\omega)^2 = 0$, and note that we may write, using equation (2.11) with $C_n(\omega) = 0$:

$$C_5^2(\omega) = 0 = \cos(5\cos^{-1}\omega)$$

and

$$5\cos^{-1}\omega = \frac{k\pi}{2}; \; k = 1, 3, 5, 7$$

from which

$$\omega = \cos\frac{k\pi}{10}$$

The zeros will occur in conjugate pairs and we shall also find that a zero occurs at $\omega = 0$:

$\omega_0 = 0 \qquad\qquad\qquad s = 0$

$\omega_1 = \cos 18° = \pm 0.951 \qquad s = \pm j0.951$

$\omega_3 = \cos 54° = \pm 0.58 \qquad s = \pm j0.58$

$N(s) = s(s^2 + 0.336)\,(s^2 + 0.904)$

$N(s) = s^5 + 1.24s^3 + 0.304s$

From which

$$\rho(s) = \frac{N(s)}{D(s)} = \frac{s^5 + 1.24s^3 + 0.304s}{s^5 + 0.572s^4 + 1.442s^3 + 0.559s^2 + 0.419s + 0.064}$$

Using $Z_{IN}(s) = \dfrac{1 + \rho(s)}{1 - \rho(s)}$, we obtain after routine manipulation:

$$Z_{IN}(s) = \frac{2s^5 + 0.522s^4 + 2.682s^3 + 0.56s^2 + 0.723s + 0.064}{0.572s^4 + 0.202s^3 + 0.56s^2 + 0.115s + 0.064}$$

Employing the synthetic division process, $Z_{IN}(s)$ reduces to:

$$Z_{IN}(s) = 3.496s + \cfrac{1}{0.79s + \cfrac{1}{4.39s + \cfrac{1}{0.734s + \cfrac{1}{3.406s + 1}{\overline{1}}}}}$$

The normalised circuit is shown in figure 6.6.

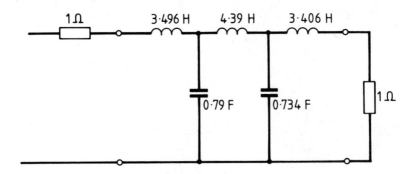

Figure 6.6   Prototype fifth-order Chebyshev filter

Scaling factors:

$$k_f = 2\pi \times 5 \times 10^3 = \pi \times 10^4$$

$$k_m = 600$$

$$C_o = \frac{C_n}{k_m k_f}$$

$$R_o = 600 \ \Omega$$

$$L_o = L_n \left(\frac{k_m}{k_f}\right)$$

The final, scaled circuit is shown in figure 6.7 which also shows the frequency response.

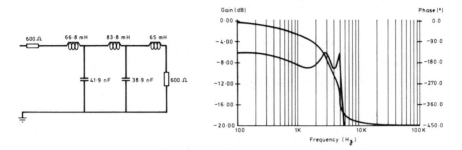

Figure 6.7   Scaled circuit with response for example 6.3

The foregoing example clearly indicates that the design of passive filters having either a Butterworth or Chebyshev response is a laborious and time-consuming exercise. Fortunately, the designer has recourse to tables wherein the normalised inductor and capacitor component values are computed for any order n of filter. The only hardship facing the designer is that of scaling the circuit to the filter specification. Of course if the reader objects to his work being done by others, then by all means reject the idea of tables!

6.4 Frequency transformations

Filter design is usually based on consideration of the prototype low-pass circuit. Other types of response are often required and these may be achieved using an appropriate frequency transformation which means that the specification of a low-pass filter may be transformed into those of any other filter using a *reactance transformation*. By means of such a transformation, the low-pass frequency variable Ω is replaced by a realisable function $X(\omega)$ associated with the variable ω. Other filter responses may be achieved by the appropriate choice of $X(\omega)$, it being noted that such a function is physically realised by using pure inductors and capacitors.

Let $\Omega = X(\omega)$ be the transformation, and

$$X(\omega) = H \frac{(\omega_2^2 - \omega^2)(\omega_4^2 - \omega^2) \dots}{(\omega_1^2 - \omega^2)(\omega_3^2 - \omega^2) \dots} \qquad (6.9)$$

The reactance function has the form shown in figure 6.8.

If we define the low-pass cut-off frequency to be $-\Omega c$ and Ωc, then the pass-band extends from $-\Omega c \leqslant \Omega \leqslant \Omega c$ which is itself transformed into a number of pass bands with centre frequencies at 0, $\pm\omega_2$, $\pm\omega_4$. The components of the transformed filters are obtained by applying the transformation to the reactance of the elements of the prototype. For example

$$\Omega L_p = X(\omega)L_p$$

$$\frac{1}{\Omega C_p} = \frac{1}{X(\omega)C_p}$$

where L_p, C_p are the prototype values.

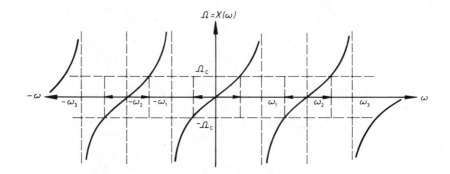

Figure 6.8   Reactance function representing $X(\omega)$

6.4.1 Low-pass to high-pass

Here we select the reactance function:

$$\Omega = -\frac{H}{\omega} \qquad\qquad (6.10)$$

This function is shown in figure 6.9.

It can be seen that using this transformation, the low-pass pass-band is transformed into two pass-bands $(-\omega_c, -\infty); (\omega_c, \infty)$.

$$j\Omega L_p \rightarrow -j\frac{H}{\omega}L_p = \frac{1}{j\omega C_p} \quad \text{where } C_p = \frac{1}{HL_p}$$

$$\frac{1}{j\Omega C_p} \rightarrow -\frac{\omega}{jHC_p} = \frac{j\omega}{HC_p} = j\omega L_p \quad \text{where } L_p = \frac{1}{HC_p}$$

The circuits are shown in figure 6.10.

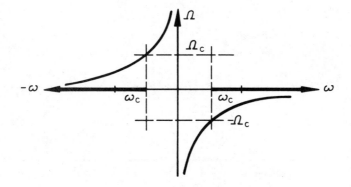

Figure 6.9   Reactance function $\Omega = -(H/\omega)$

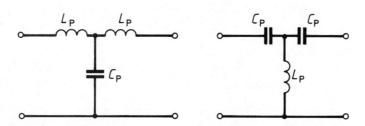

Figure 6.10   Low-pass and low-pass transformed prototype

Example 6.4

Design a high-pass network having a cut-off frequency of 5 kHz and being terminated in 600 Ω resistors.

From table 6.1 we obtain, for a normalised low pass having equal terminations:

$$L_p = 1 \text{ H}, \ C_p = 2 \text{ F}, \ \Omega_c = 1, \ H = 1$$

Using the relationships obtained:

$$C_p = \frac{1}{L_p} = 1 \text{ F}, \ L_p = \frac{1}{C_p} = 0.5 \text{ H}$$

Scaling values are

$$k_m = 600, \ k_f = \frac{2\pi \times 5 \times 10^3}{1} = 3.14 \times 10^4$$

giving for the scaled circuit:

$C = 0.053 \ \mu F, \ L = 9.55 \ mH$

The three circuits are shown in figure 6.11.

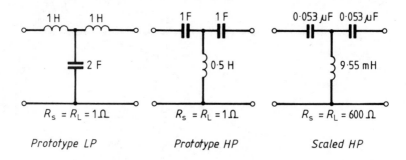

Figure 6.11   Circuits associated with example 6.4

6.4.2 Low-pass to band-pass

Here we select the reactance function to be:

$$\Omega = -A \ \frac{\omega_n^2 - \omega^2}{\omega} \tag{6.11}$$

which is shown in figure 6.12.

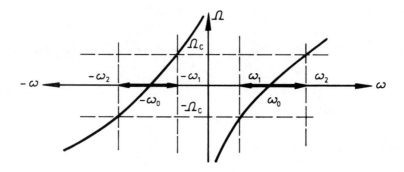

Figure 6.12   Reactance function $\Omega = -A \ ((\omega_n^2 - \omega^2)/\omega)$

The function is seen to transform the low-pass pass-band $(-\Omega_c . \Omega_c)$ into two band-pass pass-bands $(\omega_1, -\omega_2)$. The transformed frequencies are

Ω_c corresponds to $(-\omega_1, \omega_2)$

$-\Omega_c$ corresponds to $(\omega_1, -\omega_2)$

Applying the transformation to the low-pass circuit we obtain for $\omega_0^2 = \omega_1 \omega_2$:

$$j\Omega L_p = -jAL_p \frac{(\omega_0^2 - \omega^2)}{\omega} = -jAL_p \frac{\omega_0^2}{\omega} + j\omega AL_p$$

$$j\Omega L_p = j\omega AL_p + \frac{1}{\dfrac{j\omega}{AL_p\omega_0^2}} = j\omega L_{BP} + \frac{1}{j\omega C_{BP}}$$

where $L_{BP} = AL_p$, $C_{BP} = \dfrac{1}{AL_p\omega_0^2}$

Similarly:

$$\frac{1}{j\Omega C_p} = -\frac{\omega}{jC_p(\omega_0^2 - \omega^2)A} = \frac{j\omega}{AC_p(\omega_0^2 - \omega^2)} = \frac{1}{j\omega AC_p + \dfrac{AC_p\omega_0^2}{j\omega}}$$

where $A = \dfrac{\Omega_c}{\omega_2 - \omega_1} = Q$, for normalised values of frequency.

The prototype low-pass and band-pass circuits are shown in figure 6.13.

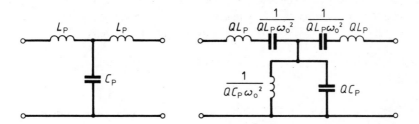

Figure 6.13   Low-pass and band-pass transformed prototypes

Example 6.5

A band-pass filter is required to have a centre frequency of 500 Hz, a bandwidth of 100 Hz and to be terminated at both ends with 1 kΩ resistors.

The normalised circuits will be those previously shown in figure 6.13 but with $\omega_0 = 1$ rad/s:

$$A = Q = \frac{f_0}{\Delta f} = \frac{500}{100} = 5$$

The scaling factors are:

$$k_m = 1000, \quad k_f = 2\pi \times 500$$

$$QL_p = 5 \text{ H}; \quad \frac{1}{QL_p} = 0.2 \text{ F}; \quad QC_p = 10 \text{ F}; \quad \frac{1}{QC_p} = 0.1 \text{ H}$$

The scaled circuit with its response is shown in figure 6.14.

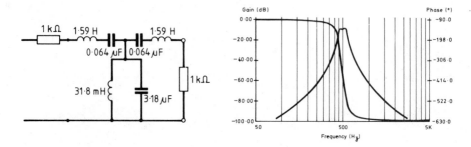

Figure 6.14   Scaled band-pass circuit and responses

We will end this chapter by noting that by means of tables of data, prototype passive filters may be designed around appropriately quoted specifications, namely Butterworth, Chebyshev, or the more exotic types based on elliptic integral theory. Furthermore, by means of an appropriately chosen frequency transformation, high-pass and band-pass circuits may be obtained.

Problems

6.1. The circuit figure 6.15 shows a cross-over network inserted between the output of an audio amplifier and woofer and tweeter loudspeakers. Using the data in table 6.1, design the prototype low- and high-pass filters and scale the circuit to give a cross-over frequency of 4 kHz with equal terminations of 8 Ω.
[*Ans.* 0.477 mH, 0.159 mH, 6.6 μF; 3.33 μF, 0.238 mH, 9.95 μF]

6.2. A Chevyshev filter is to have the following specification: pass band from 1 kHz to infinity with a permitted 1 dB of ripple; stop-band attenuation to be at least 80 dB down at 100 Hz. The circuit is to be terminated in 1 kΩ resistors at both ends.

   Using example 6.3 as a guide, obtain the complete scaled circuit, working throughout from first principles.
[*Ans.* $n = 4$, 76 nF, 0.15 H, 56 nF, 0.2 H]

6.3. A band-pass filter is required to have a centre frequency of 500 Hz, a bandwidth of 100 Hz and is to be terminated in 1 kΩ pure resistors. Design the filter using a third-order low-pass prototype and the low-pass to band-pass transformation.

[*Ans.* 1.59 H, 64 nF, 31.8 mH, 3.18 μF, 64 nF, 1.59 H]

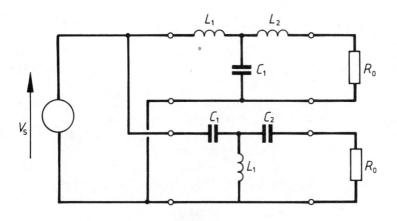

Figure 6.15

6.4. A filter circuit is required to have the following specifications: stop band to extend from dc up to 100 Hz; pass band to extend from 1 kHz up to a very high frequency. The cut-off frequency is to be 1 kHz and the filter is to be at least 200 dB down at 100 Hz; the circuit is to be terminated in 800 Ω resistors at both ends and is to be a maximally flat Butterworth design. Using the equations in chapter 2 and table 6.1, design the required filter using the low-pass to high-pass transformation and apply appropriate scaling factors. Calculate, for the order of circuit chosen, the values of stop- and pass-band attenuations.

[*Ans.* n = 10: 0.405 H, 0.22 μF, 0.09 H, 0.11 μF, 0.07 H, 0.1 μF, 0.713 μF, 0.14 μF, 0.14 H, 0.633 μF]

7 Gyrator-based Filter Circuits

7.1 Introduction

A major problem encountered by filter designers is that caused by the need for inductance at low frequencies. We will now consider a concept which will enable the designer to dispense with inductors for certain applications by using what is referred to in the literature as *the passive network simulation method*. Central to the basis of the design is a device referred to as a gyrator whose principle of operation was outlined by Tellegan. It will be seen that we fortunately possess an op-amp based gyrator circuit and several such devices are manufactured using IC technology. A distinct advantage of the active simulation approach is that the circuits produced have low sensitivities which are basically those of the passive counterparts. We shall see that a distinct disadvantage of producing *synthetic* inductance, as it is called, is that being an active device, one of its terminals must be connected to earth. Furthermore, if matched op-amps are used, then the effects of non-ideal amplifier parameters on the inductance realisation are minimised.

7.2 Properties of an ideal gyrator

The *ideal* gyrator is a four-terminal (two-port) network which presents, at either port, an input impedance which is proportional to the admittance connected across the other port. With reference to figure 7.1 it is considered that

$$Z_{11} = r^2 Y_{22} \tag{7.1}$$

likewise

$$Y_{11} = g^2 Z_{22} \tag{7.2}$$

where g is defined as the *gyration conductance:*

$$g = \frac{1}{r} \tag{7.3}$$

Observation of equations (7.1) and (7.2) suggests that if the termination is a pure resistor then the input impedance is also a pure resistor. If the terminating component is a pure capacitor, however, then the input impedance is a pure inductor and vice versa. We may conclude, therefore, that the device *inverts* the impedance connected to it. Furthermore, the *ideal* gyrator is assumed to dissipate or store no power, and is referred to as being a passive, loss-less element.

To study further the behaviour of the gyrator, we will characterise the network of figure 7.1 by its appropriate y parameters.

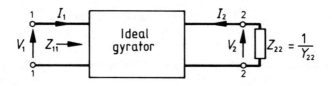

Figure 7.1 Two-port network terminated in Z_{22}

$$I_1 = y_{11} V_1 + y_{12} V_2$$
$$I_2 = y_{21} V_1 + y_{22} V_2 \qquad (7.4)$$

where y_{11} = short-circuit input admittance
y_{12} = reverse short-circuit transfer admittance
y_{21} = forward short-circuit transfer admittance
y_{22} = short-circuit output admittance.

From figure 7.1 we see that using the given polarities for currents and voltages

$$I_2 = -Y_{22} V_2 \qquad (7.5)$$

Manipulation of equations (7.4) and (7.5) yields

$$Y_{11} = \frac{I_1}{V_1} = y_{11} - \frac{y_{12} y_{21}}{y_{22} + Y_{22}} \qquad (7.6)$$

The conditions which must be imposed to enable the two-port network to act as an ideal gyrator are

$$y_{11} = 0 = y_{22}$$
$$y_{12} y_{21} = -g^2 \qquad (7.7)$$
$$Y_{11} = g^2 Z_{22}$$

Choosing $y_{12} = g$ and $y_{21} = -g$ we obtain from equation (7.4) the relationships which exist between the current at one port and the voltage at the other port in the form

$$I_1 = gV_2$$
$$I_2 = -gV_1 \qquad (7.8)$$

The symbol for an ideal gyrator is shown in figure 7.2.

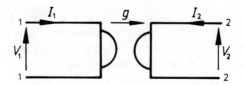

Figure 7.2   Symbol for an electrical gyrator

Earlier it was stated that a property of a gyrator was that of having the ability to invert impedances. Consider that impedance Z_{22} is a pure capacitor $Z(j\omega) = 1/j\omega C$. Using equation (7.1) produces

$$Z_{11} = r^2 j\omega C = j\omega(r^2 C)$$

Here it can be seen that the capacitor C F has been *inverted* into an inductor $L = r^2 C$ H.

A circuit which may be considered as a practical realisation of an ideal gyrator is shown in figure 7.3.

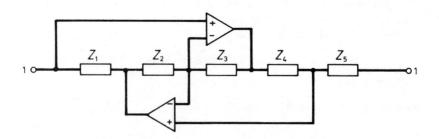

Figure 7.3   Two op-amp realisation of a gyrator

7.3 The generalised impedance converter – GIC

The circuit which is shown in figure 7.3 may be re-drawn as shown in figure 7.4 where the impedance Z_5 has been removed from outside the circuit enclosed within the box and which is referred to as a GIC.

If the circuit of figure 7.4 is analysed for its input impedance it can be shown that

$$Z_{11} = \frac{V_1}{I_1} = \frac{Z_1 Z_3 Z_5}{Z_2 Z_4} \tag{7.9}$$

Further, if the element Z_5 is removed and terminals 2–2 created in its place, and an element Z_0 connected across terminals 1–1, it can also be shown that

$$Z_{22} = \frac{V_2}{I_2} = \frac{Z_2 Z_4 Z_0}{Z_1 Z_3} \qquad (7.10)$$

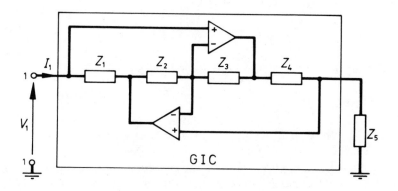

Figure 7.4 Circuit drawn as a GIC

The new circuit is shown in figure 7.5.

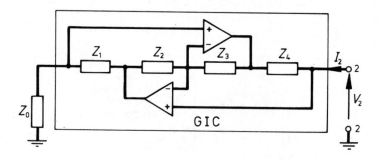

Figure 7.5 Circuit drawn as a GIC – termination Z_0

The GIC circuits of figures 7.4 and 7.5 are mainly due to Antoniou, and we shall see that both circuits have an important role to play in the simulation of so-called *synthetic inductors* associated with active realisations of passive circuits.

Consider the circuit shown in figure 7.4 and let element Z_4 be a capacitor with the remaining elements being pure resistors. Using equation (7.9) we find that

$$Z_{11} = \frac{R_1 R_3 R_5 C_4 s}{R_2} = L_{eq} s \qquad (7.11)$$

It is seen that the GIC *converts* a resistor R_5 into an equivalent inductor having the value given by equation (7.11). Similarly, using the circuit of figure 7.5 and equation (7.10) with Z_0 and Z_4 capacitors, the remaining elements being pure resistors, we find that

$$Z_{22} = \frac{R_2}{R_1 R_3 C_4 C_0 s^2} = \frac{1}{D s^2}$$

Also noting that $s = j\omega$, then the expression becomes

$$Z(j\omega) = -\frac{1}{D\omega^2} \tag{7.12}$$

Here we can see that the input impedance is negative and that it also varies with frequency. It is therefore referred to as a *frequency dependent negative resistor* (FDNR). The concept was initially introduced by Bruton who also introduced a method of FDNR implementation for the simulation of ladder filters. He also introduced the notion of scaling elements by a factor $(1/s)$ whereby resistors became inductors, inductors became resistors and capacitors became FDNRs. Using this idea it can be seen that by using such a factor, inductors are no longer described. The principle is outlined below and summarised in figure 7.6 where the symbol for an FDNR should be noted. It should also be noted that the Bruton transformed network should have an identical transfer function to the original network being investigated.

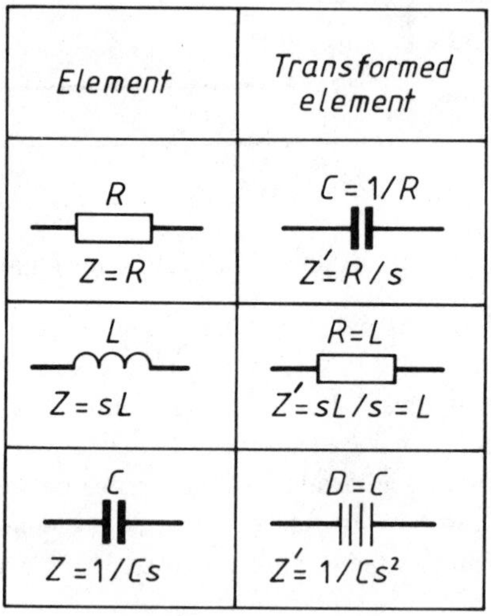

Element	Transformed element
R ▭ $Z = R$	$C = 1/R$ ⊣⊢ $Z' = R/s$
L ⌇⌇⌇ $Z = sL$	$R = L$ ▭ $Z' = sL/s = L$
C ⊣⊢ $Z = 1/Cs$	$D = C$ ⊣⊦⊢ $Z' = 1/Cs^2$

Figure 7.6 Transformed elements

Gyrator module integrated circuits are produced; one such being manufactured by National Semiconductors is shown in figure 7.7.

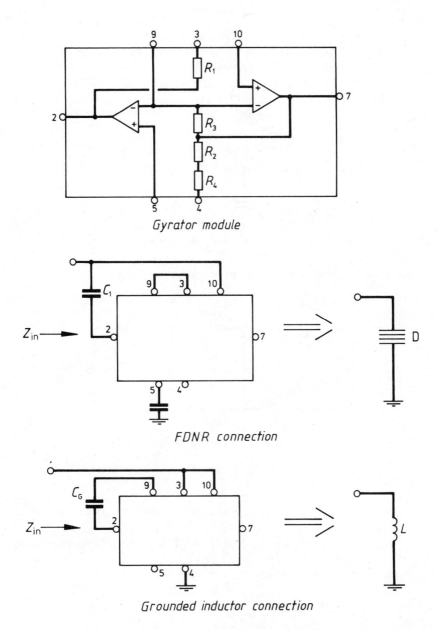

Gyrator module

FDNR connection

Grounded inductor connection

Figure 7.7 Gyrator module schematic

Two examples will now be considered with which to illustrate the principles outlined in this chapter.

Example 7.1

A low pass Butterworth filter is to be designed using FDNRs. The cut-off frequency is to be 1500 Hz and the circuit is to be of the third order.

Using the data from table 6.1 from chapter 6, we obtain the normalised low-pass passive prototype shown in figure 7.8. Employing the Bruton transformation to each element, we obtain the circuit shown in figure 7.9.

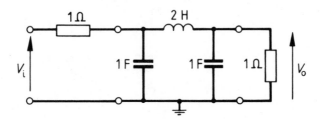

Figure 7.8   Prototype low-pass circuit

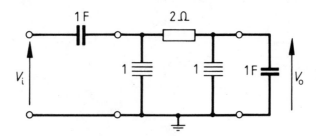

Figure 7.9   Bruton transformation of figure 7.8

Next we convert figure 7.9 into the final normalised circuit incorporating the FDNRs; this is shown in figure 7.10.

Finally, scaling is applied to produce the final scaled circuit shown in figure 7.11, with the response shown in figure 7.12.

We select $C_0 = 20$ nF which gives

$$k_f = 2\pi \times 1500 = 3000\pi$$

$$k_m = \frac{1}{3000\pi \times 20 \times 10^{-9}} = 5305$$

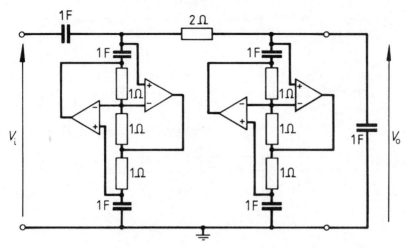

Figure 7.10 Gyrator-based prototype

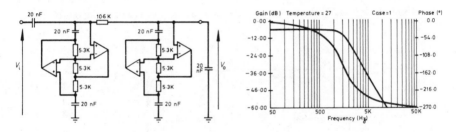

Figure 7.11 Scaled circuit   Figure 7.12 Circuit frequency responses

Example 7.2

We are required to design a fifth-order Butterworth high-pass filter having a cut-off frequency of 100 Hz and terminated in 600 Ω resistors.

Using table 6.1 from chapter 6, we obtain the prototype low-pass circuit shown in figure 7.13.

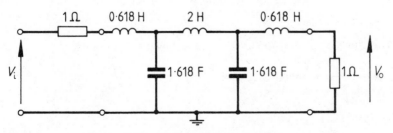

Figure 7.13 Prototype T-section low-pass

Using the low-pass to high-pass transformation, the prototype high-pass equivalent circuit is obtained and is shown in figure 7.14.

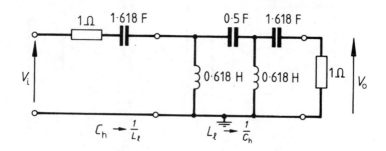

Figure 7.14   Prototype high-pass circuit

Clearly, it can be seen that we shall require a GIC network to simulate the 0.618 H inductors. The prototype gyrator-based circuit is shown in figure 7.15.

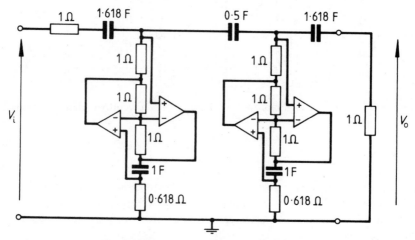

Figure 7.15   Prototype GIC-based circuit

After applying scaling, the final circuit is shown in figure 7.16.

$$k_f = 200\pi, \quad k_m = 600, \quad k_m k_f = \frac{C_n}{C_o}$$

and

$$C_o = \frac{1}{200\pi \times 600} = 2.65 \ \mu F$$

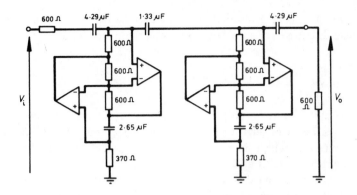

Figure 7.16   Scaled gyrator circuit

The frequency response of the circuit is shown in figure 7.17.

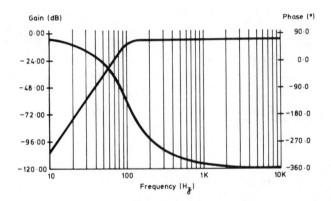

Figure 7.17

From a consideration of the two examples discussed, a number of points emerge which are relevant to the correct realisation of gyrator-based circuits.

1. Synthetic inductors should be tested before being connected into the circuit. This may be achieved by resonating the inductor with a known capacitor.
2. Because the FDNR is an active element, one end of it must always be connected to earth. Low-pass filters satisfy this requirement since application of the $RLC:CRD$ transformation transforms the prototype passive, earthed capacitor into an earthed FDNR.
3. The use of matched op-amps reduces the errors resulting from non-ideal op-amp parameters on the simulated components.
4. Floating synthetic inductors can be made, but their performance is not as good as those which may be grounded.

5. Attention should be given within the gyrator design to providing paths for the small dc currents and offset voltages at each op-amp input.

7.4 Conclusions

We have seen that it is possible to simulate passive filter circuits by means of active circuits. The main advantage to be gained is that inductors are not involved in the simulation, which implies a physically smaller, perhaps cheaper and certainly more accurate filter design.

The sensitivities of the simulation are effectively those of the passive circuit which are themselves quite low. Also, band-pass circuits may be realised using the basic GIC circuits, but having modified terminating components. The technique is beyond the intended scope of this book, but suffice it to say that we may realise certain complex impedances by considering the terminations to be a combination of resistors and capacitors.

For the present, gyrator filters have been integrated on to a single chip for use as a video filter within a TV receiver and have been found to be suitable for mass production. Another example is that of a twin band-reject filter which was required to interface with an existing speech communication system in order to suppress two-tone frequencies. There is no doubt that gyrator-based filter circuits have a considerable future and should not be regarded purely as a 'novelty' concept.

Problems

7.1. The circuits shown in figure 7.18 are those of figures 7.4 and 7.5 re-drawn into a different circuit configuration. Derive expressions for Z_{11} and Z_{22} and compare the results with equations (7.9) and (7.10) in the text.

7.2. Using the data for a Butterworth filter from chapter 2, design a fourth-order filter using the techniques outlined in chapter 7 to have the following specification. Pass band to extend from dc up to 1 kHz with a maximum attenuation of 3 dB. Stop-band attenuation to be 80 dB down at 10 kHz. Use 0.01 μF capacitors in the design.
   [*Ans.* The two FDNRs will contain 16 kΩ, 29 kΩ and 0.01 μF]

7.3. Using the prototype circuits obtained in example 6.3, obtain the gyrator circuit form to give the following specification: pass-band 1 kHz up to infinity with 1 dB of ripple; stop-band attenuation to be better than 80 dB at 100 Hz. Use a magnitude scaling factor of 1000.
   [*Ans.* FDNR1: 1 kΩ, 1 kΩ, 1 kΩ, 159 nF, 940 Ω.
        FDNR2: 1 kΩ, 1 kΩ, 1 kΩ, 159 nF, 1.27 kΩ]

7.4. The circuit shown in figure 7.19 is often attributed to Fliege and possesses low sensitivity. Using nodal analysis and the assumption that the op-amps are ideal, show that

$$\frac{V_o}{V_i} = \frac{Y_A(Y_1 Y_3 - Y_0 Y_2) + Y_B Y_2 (Y_4 + Y_5)}{Y_1 Y_3 (Y_A + Y_5) + Y_2 Y_4 (Y_B + Y_0)}$$

Further show that a low-pass filter is obtained when

$$Y_1 = \frac{1}{R_1} + j\omega C_1, \quad Y_2 = \frac{1}{R_2}, \quad Y_3 = j\omega C_3, \quad Y_4 = \frac{1}{R_4}, \quad Y_5 = \frac{1}{R_5}, \quad Y_A = 0,$$

$$Y_B = \frac{1}{R_B} \quad Y_0$$

What would be the conditions for a band-pass filter?

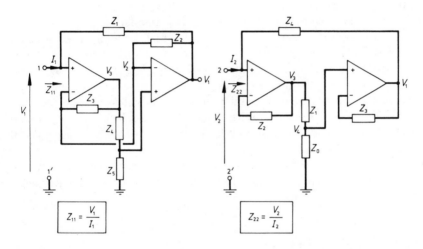

Figure 7.18

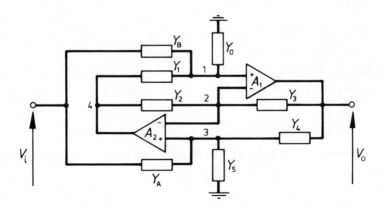

Figure 7.19

8 Sensitivity

8.1 Introduction

Circuits or networks are constructed to produce a specified input/output relationship and since the solution to the design problem may not be unique, a variety of circuits may be constructed to realise the same transfer function.

For so long as ideal elements are used under ideal conditions, then one circuit may perform as well as any other. In practice, as we have seen, certain circuits may perform better than others because they are less sensitive to variation in component values. Furthermore, these circuits may be no more expensive to construct and it is therefore useful to have a quantitative measure, whereby a comparison may be made between circuits when the components vary from the ideal values because of tolerances related to their manufacture, aging and the effects of humidity and temperature.

Sensitivity functions are used for this purpose and they involve a study of the way in which the circuit transfer function changes when changes are made to individual component values.

8.2 Definition of sensitivity

Consider a circuit in which a change of 10 per cent is made in one of the circuit components and assume that this change produces a change in the output of the circuit of 50 per cent. The sensitivity of the circuit could be defined as:

$$S_x^H = \frac{\text{change in output}}{\text{change in input}} = \frac{50}{10} = 5$$

Therefore the sensitivity of the circuit transfer function H to a change in a component x is 5. We may define the sensitivity function S_x^H in the following way:

$$S_x^H = \frac{x}{H} \frac{\partial H}{\partial x} \qquad (8.1)$$

In circuit analysis the transfer function $H(j\omega)$ is usually expressed as the quotient of two frequency dependent variables:

$$H(j\omega) = \frac{N(j\omega)}{D(j\omega)}$$

142

which on differentiation yields:

$$\frac{\partial H}{\partial x} = \frac{N'D - D'N}{D^2} \tag{8.2}$$

where N', D' are the partial derivatives of N and D with respect to x. By combining equations (8.1) and (8.2) we obtain:

$$S_x^H = \frac{x}{H} \frac{\partial H}{\partial x} = \frac{x}{H} \left[\frac{N'D - D'N}{D^2} \right] \quad \text{and} \quad H = \frac{N}{D}$$

which finally produces the expression:

$$S_x^H = x \left[\frac{N'}{N} - \frac{D'}{D} \right] \tag{8.3}$$

It was stated earlier that the transfer function is usually written as

$$H(j\omega) = \frac{N(j\omega)}{D(j\omega)} = |H(j\omega)| \exp(j\phi)$$

Substituting this expression into (8.1) yields:

$$S_x^H = \frac{x}{|H(j\omega)|} \frac{\partial}{\partial x} |H(j\omega)| + j\phi \left(\frac{x}{\phi} \frac{\partial \phi}{\partial x} \right) \tag{8.4}$$

and

$$S_x^H = S_x^{|H|} + j\phi S_x^\phi$$

The *real* part of this expression represents the magnitude sensitivity of $H(j\omega)$

with respect to $x(S_x^{|H(j\omega)|})$. The *imaginary* part represents the phase sensitivity with respect to $x(S_x^\phi)$. It should be noted that angle ϕ must be in radians. Also note that to obtain S_x^ϕ, the imaginary part of the expression must be divided by ϕ radians.

8.3 Sensitivity of the VCVS filter

From chapter 4 we re-call the transfer function for the filter, namely

$$H(j\omega) = \frac{V_o}{V_i}(j\omega) = \frac{\dfrac{1}{R_1 R_2} K}{-\omega^2 C_1 C_2 + j\omega \left(\dfrac{C_1}{R_1} + \dfrac{C_1}{R_2} + \dfrac{C_2}{R_2} - \dfrac{C_2 K}{R_2} \right) + \dfrac{1}{R_1 R_2}}$$

Consider now the sensitivity of the circuit transfer function for variation of capacitor C_1; using equation (8.3) we obtain

$$S_{C_1}^H = C_1 \left[\frac{N'}{N} - \frac{D'}{D} \right], \quad \text{where } N' = \frac{\partial N}{\partial C_1}, \; D' = \frac{\partial D}{\partial C_1}$$

$$S_{C_1}^H = C_1 \left[\frac{0}{N} - \frac{\left\{ -\omega^2 C_2 + j\omega \left(\frac{1}{R_1} + \frac{1}{R_2} \right) \right\}}{D} \right]$$

$$S_{C_1}^H = \frac{\omega^2 C_1 C_2 - j\omega C_1 \left(\frac{1}{R_1} + \frac{1}{R_2} \right)}{D}$$

Similarly for the remaining components:

$$S_{C_2}^H = \frac{\omega^2 C_1 C_2 - j\omega C_2 \dfrac{(1 - K)}{R_2}}{D}$$

(8.5)

$$S_{R_1}^H = \frac{\omega^2 C_1 C_2 - j\omega \left\{ \dfrac{C_1}{R_2} + \dfrac{C_2}{R_2} (1 - K) \right\}}{D}$$

$$S_{R_2}^H = \frac{\omega^2 C_1 C_2 - j\omega \dfrac{C_1}{R_1}}{D}$$

where

$$D = -\omega^2 C_1 C_2 + j\omega \left\{ \frac{C_1}{R_1} + \frac{C_1}{R_2} + \frac{C_2}{R_2} (1 - K) \right\} + \frac{1}{R_1 R_2}$$

It can be seen from the expressions that the circuit sensitivity for each component varies with frequency and is not constant. To illustrate how sensitivity analysis may be used to predict changes in gain and phase caused by changes in component values, a practical example will now be considered.

Example 8.1

A VCVS filter design is shown in figure 8.1 which is to have a pass-band gain of 20 dB and a cut-off frequency of 10 kHz. The gain and phase responses are shown in figure 8.2 with the response to a -10 per cent change in C_1 also shown.

Finally, the magnitude and phase sensitivities are shown in figure 8.3 and values calculated using equations (8.5).

We wish to calculate:

(a) the change in gain at 6 kHz and
(b) the change in phase at 3 kHz when the capacitor value is decreased by 10 per cent.

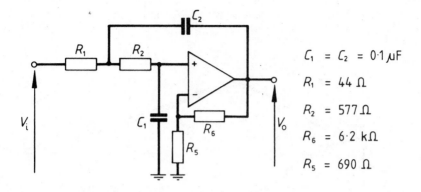

Figure 8.1   VCVS low-pass filter

$$C_1 = C_2 = 0.1 \mu F$$
$$R_1 = 44 \, \Omega$$
$$R_2 = 577 \, \Omega$$
$$R_6 = 6.2 \, k\Omega$$
$$R_5 = 690 \, \Omega$$

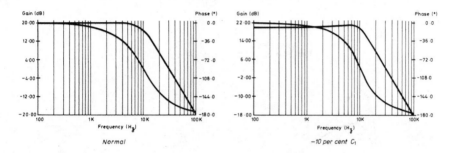

Figure 8.2   Gain/phase

Insertion of the component values into equation (8.5) gives the following sensitivities:

$$S_{R_1}^H = \frac{\omega^2 + j1.4 \times 10^5 \, \omega}{D}$$

$$S_{R_2}^H = \frac{\omega^2 - j2.23 \times 10^5 \, \omega}{D}$$

$$S_{C_1}^H = \frac{\omega^2 - j2.45 \times 10^5 \, \omega}{D}$$

$$S_{C_2}^H = \frac{\omega^2 + j1.56 \times 10^5 \, \omega}{D}$$

where

$$D = 3.94 \times 10^9 - \omega^2 + j8.86 \times 10^4 \, \omega$$

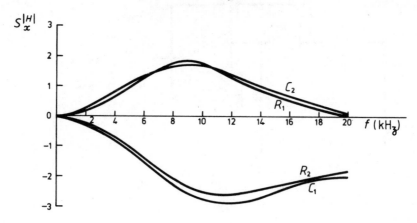

Magnitude sensitivity for VCVS filter

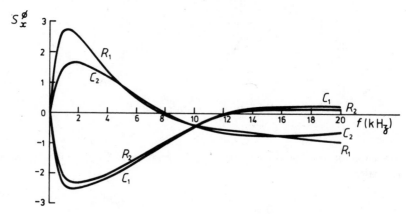

Phase sensitivity for VCVS filter

Figure 8.3   Magnitude/phase sensitivities

The next step is to consider a range of values for ω and obtain the component sensitivities for each value of ω. Finally, magnitude and phase sensitivity graphs are plotted as shown in figure 8.3. To illustrate the method of obtaining the sensitivities, we will consider the conditions for a frequency of 5 kHz.

$$S_{R_1}^H = \frac{(2\pi \times 5 \times 10^3)^2 + j1.4 \times 2\pi \times 5 \times 10^8}{(3.94 \times 10^9 - (2\pi \times 5 \times 10^3)^2) + j8.86 \times 2\pi \times 5 \times 10^7} = 1.11\angle 34.2°$$

$$S_{R_1}^H = 0.92 + j0.62 \text{ which on using (8.4) gives}$$

$$S_{R_1}^{|H|} = 0.92 \quad S_{R_1}^{\phi} = \frac{0.62}{\phi} = \frac{0.62}{\pi/4} = 0.8$$

where ϕ is the phase obtained from the *normal* response characteristic of figure 8.2 at 5 kHz.

From equation (8.4) we may obtain the incremental change in response in the form:

$$\partial H \approx |H| \, S_x^{|H|} \, \frac{\partial x}{x} \tag{8.6}$$

$$\partial \phi \approx \phi \, S_x^{\phi} \, \frac{\partial x}{x}$$

(a) From figure 8.2 we see that the gain at 6 kHz is 21 dB and also from figure 8.2 the designed gain is 19.5 dB.

$$|H| = \text{Antilog}_{10} \, \frac{19.5}{20} = 9.44$$

$$\frac{\partial C_1}{C_1} = -\frac{0.01}{0.1} = -0.1 \quad \text{and from figure 8.3}$$

$$S_{C_1}^{|H|} = -1.5$$

Therefore

$$\partial H \approx 9.44 \times -1.5 \times -0.1 = 1.416 \quad \text{giving}$$

$$|\hat{H}| = 9.44 + 1.416 = 10.86 \quad \text{or} \quad 20.8 \text{ dB}$$

This is good agreement with the value of 21 dB.

(b) Also from figure 8.2 we see that at $f = 3$ kHz, $\phi = 26°$ and from figure 8.3 $S_{C_1}^{\phi} = -2.35$.

$$\partial \phi = 26 \times -2.35 \times -0.1 = 6°$$

giving $\hat{\phi} = -26° + 6° = -20°$ which gives good agreement with $-21°$ from figure 8 2.

The two sensitivity plots show that the magnitudes $S_{C_2}^{|H|}$ and $S_{R_1}^{|H|}$ are both positive and have their largest values over a frequency range 5-14 kHz. Over this range of frequency we would expect the most marked effect on the gain response and would anticipate an increase in the gain. The converse is true for $S_{C_1}^{|H|}$ and $S_{R_1}^{|H|}$ where a reduction in gain is anticipated.

The phase sensitivity characteristics show a low negative value around 10 kHz; this suggests that within this frequency range the phase response should show very small deviation from the normal.

A further point to note from the two graphs is that the resistors (R_1, R_2) and capacitors (C_1, C_2) should be chosen to have identical temperature coefficients.

8.4 Sensitivity of K, Q and ω_0

Sensitivity analysis may also be applied to ascertain the variation of Q and ω_0 due to component variation within the circuit.

To illustrate the technique we will use the VCVS circuit of chapter 4, in which was stated the voltage transfer function:

$$\frac{V_o}{V_i}(j\omega) = \frac{\dfrac{1}{C_1 C_2 R_1 R_2} K}{-\omega^2 + j\omega \left\{ \dfrac{1}{C_2 R_1} + \dfrac{1}{C_2 R_2} + \dfrac{1}{C_1 R_2} - \dfrac{K}{C_1 R_2} \right\} + \dfrac{1}{C_1 C_2 R_1 R_2}}$$

which may be conveniently re-written as

$$\frac{V_o}{V_i}(j\omega) = \frac{K\omega_0^2}{-\omega^2 + j\omega \left(\dfrac{\omega_0}{Q}\right) + \omega_0^2} \tag{8.7}$$

where

$$K = 1 + \frac{R_6}{R_5}, \quad \omega_0 = \frac{1}{\sqrt{(C_1 C_2 R_1 R_2)}}, \quad \frac{\omega_0}{Q} = \frac{1}{C_2 R_1} + \frac{1}{C_2 R_2} + \frac{1}{C_1 R_2}(1 - K)$$

For convenience, let $P = \dfrac{\omega_0}{Q}$ and from equation (8.1):

$$S_x^{\omega_0} = \frac{x}{\omega_0} \frac{\partial \omega_0}{\partial x}, \quad S_x^Q = S_x^{\omega_0} - S_x^P \tag{8.8}$$

If we now apply equations (8.8) for each component of the filter, we obtain

$$S_{R_1}^{\omega_0} = \frac{R_1}{\omega_0} \frac{\partial \omega_0}{\partial R_1} = -\frac{R_1 R_1^{-1}(C_1 C_2 R_1 R_2)^{-\frac{1}{2}}(C_1 C_2 R_1 R_2)^{\frac{1}{2}}}{2} = -\frac{1}{2}$$

Similarly for resistor R_2 and capacitors C_1 and C_2. Consider now variation of Q with each component.

$$S_{R_1}^P = \frac{R_1}{P} \frac{\partial P}{\partial R_1} = \frac{R_1}{P} \left(-\frac{R_1^{-2}}{C_2}\right) = -\frac{1}{C_2 R_1 P}, \text{ also}$$

$$P = \frac{\omega_0}{Q}, \quad \text{giving } S_{R_1}^P = -\frac{Q}{\omega_0 C_2 R_1} = \sqrt{\left(\frac{C_1 R_2}{C_2 R_1}\right)} Q$$

Using now the previous result for $S_{R_1}^{\omega_0}$ and combining with the result for $S_{R_1}^P$:

$$S_{R_1}^Q = S_{R_1}^{\omega_0} - S_{R_1}^P = -\frac{1}{2} + \sqrt{\left(\frac{C_1 R_2}{C_2 R_1}\right)} Q \tag{8.9}$$

A similar analysis will yield the sensitivities due to the remaining components.

Sensitivity values for the gain resistors may be found as follows by re-calling the expressions:

$$P = \frac{1}{C_2 R_1} + \frac{1}{C_2 R_2} + \frac{1}{C_1 R_2}(1-K), \quad (1-K) = -\frac{R_6}{R_5}$$

$$P = \frac{1}{C_2 R_1} + \frac{1}{C_2 R_2} - \frac{R_6}{C_1 C_2 R_5}, \quad \text{also}$$

$$S_K^Q = S_K^{\omega_0} - S_K^P \quad \text{and} \quad S_K^{\omega_0} = 0, \quad \text{therefore}$$

$$S_K^P = \frac{K}{P}\frac{\partial P}{\partial K} = -\frac{K}{P} = -\frac{KQ}{\omega_0} \quad \text{and}$$

$$S_K^Q = \frac{QK}{C_1 R_2}\sqrt{(C_1 C_2 R_1 R_2)} = KQ\sqrt{\left(\frac{C_2 R_1}{C_1 R_2}\right)} \tag{8.10}$$

We may also obtain the sensitivities of gain and Q-factor for variation of the gain adjustment resistors:

$$S_{R_6}^Q = S_{R_6}^{\omega_0} - S_{R_6}^P = 0 - \frac{R_6}{C_1 R_2 R_5 P} = -Q\frac{R_6}{R_5}\sqrt{\left(\frac{C_2 R_1}{C_1 R_2}\right)}$$

$$= (K-1)Q\sqrt{\left(\frac{C_2 R_1}{C_1 R_2}\right)} \tag{8.11}$$

Similarly it can be shown that for variation of R_5 and R_6:

$$\left.\begin{array}{l} S_{R_5}^Q = -S_{R_6}^Q = -(K-1)Q\sqrt{\left(\frac{C_2 R_1}{C_1 R_2}\right)} \\[3mm] S_{R_6}^K = \frac{K-1}{K} = -S_{R_5}^K \end{array}\right\} \tag{8.12}$$

It is convenient to take as many of the components equal to unity as is possible. Let us make $R_1 = R_2 = 1\ \Omega$, $C = C_2 = 1\ \text{F}$ and resistor $R_5 = 1\ \Omega$. To obtain the value for R_6 we use equation (8.7) and with $\omega_0 = 1\ \text{rad/s}$ to yield the following relationship:

$$3 - K = \frac{1}{Q} \quad \text{and} \quad K = 1 + \frac{R_6}{R_5}$$

from which we obtain the value for R_6:

$$R_6 = 2 - \frac{1}{Q}$$

Imposing these conditions keeps the spread of component values as small as possible.

The sensitivity values shown in table 8.1 may now be constructed showing the sensitivity for each component for the condition $R_1 = R_2 = 1\ \Omega, C_1 = C_2 = 1\ F$, $R_5 = 1\ \Omega, R_6 = 2 - \dfrac{1}{Q}\ \Omega$,

Table 8.1

Component	Sensitivity		
R_1	$S^Q_{R_1} = -\dfrac{1}{2} + Q;$	$S^{\omega_0}_{R_1} = -\dfrac{1}{2}$	
R_2	$S^Q_{R_2} = \dfrac{1}{2} - Q;$	$S^{\omega_0}_{R_2} = -\dfrac{1}{2}$	
C_1	$S^Q_{C_1} = \dfrac{1}{2} - 2Q;$	$S^{\omega_0}_{C_1} = -\dfrac{1}{2}$	
C_2	$S^Q_{C_2} = -\dfrac{1}{2} + 2Q;$	$S^{\omega_0}_{C_2} = -\dfrac{1}{2}$	
R_5	$S^Q_{R_5} = 1 - 2Q$		
R_6	$S^Q_{R_6} = 2Q - 1$		
K	$S^K_{R_5} = -\dfrac{2Q - 1}{3Q - 1}$		$S^Q_K = 3Q - 1$
	$S^K_{R_6} = \dfrac{2Q - 1}{3Q - 1}$		

To illustrate only the usefulness of sensitivity values, consider for the circuit conditions outlined above that a certain design requires $Q = 10$.

Insertion of this value yields the following sensitivity values:

$$S^Q_{R_1} = 9.5,\quad S^Q_{R_2} = -9.5,\quad S^Q_{C_1} = -19.5,\quad S^Q_{C_2} = 19.5,\quad S^Q_K = 29,$$

$$S^Q_{R_5} = -19,\quad S^Q_{R_6} = 19,\quad S^K_{R_5} = -\dfrac{19}{29} = -S^K_{R_6}$$

This circuit realisation has the advantage that all the frequency selective components are easily chosen; the resistors are equal, as are also the capacitors. This makes the scaling of the circuit much easier to implement. The price to be paid is that all of the sensitivities are high, especially ($S^Q_K = 29$). The implication is that care must be taken in adjusting the gain K to obtain the specified Q-value. The high Q sensitivity with respect to the gain is certainly a disadvantage of the VCVS circuit as we saw in chapter 4. A simple example will suffice to highlight this problem.

Example 8.2

Using table 8.1, calculate the percentage change in Q and K for a ± 5 per cent change in the values of the gain adjustment resistors. Consider the circuit has $Q = 10$.

The basic sensitivity expression is obtained from equation (8.1):

$$S^Q_{R_s} = \frac{Q}{R_5} \; \frac{\partial Q}{\partial R_5} \quad \text{etc. from which}$$

$$\partial Q = S^Q_{R_s} \times \frac{R_5}{Q} \, \partial R_5$$

$$= -19 \times \frac{1}{10} \times \pm 5 = \pm 9.5 \text{ per cent} \quad \text{similarly}$$

$$\partial K = S^K_{R_s} \times \frac{R_5}{K} \times \partial R_5 = - \frac{19}{29} \times \frac{1}{\left(3 - \dfrac{1}{10}\right)} \times \pm 5 \text{ per cent} = 1.13 \text{ per cent}$$

$$\partial Q = \frac{29 \times \left(3 - \dfrac{1}{10}\right)}{10} \times \pm 1.13 = \pm 9.5 \text{ per cent}$$

The calculations show that for a ± 5 per cent change in tolerance value of the gain adjustment resistors there is a ± 1.13 per cent change in the gain and a ± 9.5 per cent change in the Q-value.

8.5 Conclusions

To conclude this chapter we should remind ourselves that sensitivity analysis of circuits is an integral part of any analogue design, be it a filter or any other circuit. Moreover, in the final analysis we are interested in minimising the deviation of the filter response due to the incremental variation of some parameter within the circuit. For example, in a highly selective (high Q) circuit the sensitivity of the poles is an important factor to consider when studying the stability of the circuit. It is only by considering the relative sensitivities of the filter circuits available that a correct choice may then be made for a particular application.

The deviation of ω_0 and Q for a particular filter, for example, may be reduced by using an op-amp having a superior GB product. This may be fine for a single design, but commercial success depends on mass-production techniques and therefore we find the op-amp derivatives of the 741 still being used. We have also discussed in chapter 1 the need to have quality components and it was mentioned that precision components could be manufactured using thin film techniques. It is also common practice to minimise the spread of component values, especially capacitance, since the value of the total capacitance determines the area of sub-strate used in hybrid integrated-circuit form electrical filters.

Problems

8.1. Using the general circuit for the MFB filter from figure 4.21 and the sensitivity technique outlined in chapter 8, obtain the component sensitivities for the circuit and compare with those for the VCVS circuit.

8.2. Obtain the sensitivities for the two circuits shown using the sensitivity equations derived in this section and compare with the values given in table 8.1.

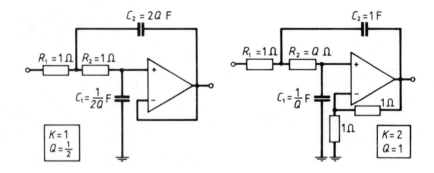

Figure 8.4

8.3. Using the graphs of figures 8.2, 8.3 etc. obtain the changes in gain and phase for the circuit of figure 8.1 for the following component variations:

(a) +10 per cent increase in R_2 at 6 kHz
(b) −10 per cent decrease in R_2 at 5 kHz
(c) ±10 per cent change in R_1 at 9 kHz
(d) +10 per cent increase in C_1 at 1 kHz
(e) −10 per cent decrease in C_1 at 10 kHz
(f) +10 per cent increase in C_2 at 5 kHz
(g) −10 per cent decrease in C_2 at 7 kHz.

Appendix: The Use of Computer Aided Design Methods

A.1 Introduction

This brief appendix deals with the formulations and methods that can be implemented for an ever-increasing number of design situations where the classical synthesis method (analytical or numerical) is considered to be inappropriate.

The modern approach to the problem is to use efficient, iterative, automatic optimisation methods to achieve a design which meets or may even exceed certain requirements. As soon as such realities as active devices, parasitic effects, non-linearities, high-frequency operation etc. are to be considered, then the classical method can, at best, only give an approximate solution to the design problem.

Generally, we could argue against the derivation of a computer algorithm for a specific filter design, since criteria such as how often the algorithm may be used and how much time is taken in its development must be overriding factors.

There are a number of CAD packages available to the designer; two of the most popular with relative novices are SPICE and MICROCAP. Quite a number of the filter circuits contained within the book were initially checked for correctness of response by means of the CAD facility. It must be stressed here that the formalities of the design were performed as in the preceding chapters and not by a computer algorithm.

A.2 The CAD approach – the MICROCAP way

The author is indebted to Spectrum Software (Sunnyvale, California, USA) for permission to include this brief overview of the MICROCAP approach to analogue circuit design.

MICROCAP is an interactive analogue circuit design and analysis tool which functions as shown in figure A.1. It allows the user to draw a circuit diagram on a CRT and automatically creates a netlist which is suitable for simulation or analysis directly from the diagram that has been drawn.

The diagrams are drawn using familiar components that are stored within a library of device models; for example, op-amps, diodes, resistors, transformers, switches, transistors (MOSFET and bipolar), etc. Running the program allows a choice from four analytical modes – ac, dc, transient and Fourier.

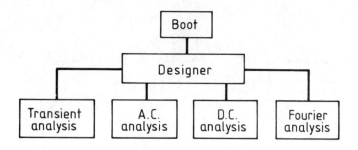

Figure A.1

Finally, the circuit diagram and the analysis may be dumped to a printer in graphic or tabular form. Alternatively, a plotter may be used to display the recorded and stored data.

Simply in order to illustrate the use of a CAD tool such as MICROCAP, we will consider a simple design example. In section A.3 we will further consider a laboratory exercise to obtain the response of the chosen design.

The design specification is that a band-pass filter is required, having a bandwidth of 500 Hz, a centre frequency of 1 kHz and a centre frequency gain of 20 dB. The actual filter design is covered in chapter 4, example 4.7, and the component values are listed as:

$$R_1 = 3.14 \text{ k}\Omega, \ R_2 = 15.9 \text{ k}\Omega, \ R_3 = 47.7 \text{ k}\Omega, \ C_1 = 0.01 \ \mu\text{F}, \ C_2 = 0.02 \ \mu\text{F}$$

The circuit is drawn as shown in figure A.2, and the ac analysis mode is selected. This initially attributes node numbers to nodal points within the circuit. A prologue of condition is then presented to the user, as shown in table A.1; it is at this point that the user must insert the appropriate data relevant to the design, in order for a correct analysis of the circuit to be achieved. For example, in the design under consideration, the input node is 1 and the output node is 5. Notice also that the number of cases taken is 3. This facility permits the use of Worst-Case Analysis procedures based on the component tolerances. On acknowledging 'Yes' to the final prompt, the program will be run by connecting a constant-amplitude, variable-frequency source to the input node, the response being displayed on the CRT screen and shown in figure A.3.

A further prompt will ask whether another RUN is required, whether to DUMP the results to a printer, or if the user wants to return to the circuit in order to make possible alterations. Clearly, it can be seen that even by using a relatively unsophisticated CAD technique, a considerable time-saving facility is available to the designer.

A typical library of devices is shown in figure A.4, along with a specification for the op-amp used in the analysis.

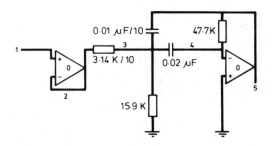

Figure A.2

Table A.1 Prologue values per ac analysis

	Analysis limits
Lowest frequency	100
Highest frequency	1E4
Lowest gain (dB)	−8
Highest gain (dB)	22
Lowest phase shift	−270
Highest phase shift	−90
Lowest group delay	1E-9
Highest group delay	1E-4
Input node number	1
Output node number	5
Minimum accuracy (%)	5
Auto or Fixed frequency step (A, F)	A
Temperature (Low/High/Step)	27
Number of cases	3
Output: Disk, Printer, None (D, P, N)	N
Save, Retrieve, 'Normal run (S, R, N)'	N
Default plotting parameters (Y, N)	Y
Are these correct (Y, N)	Y

A.3 The laboratory procedure

The circuit was tested in the laboratory using the set-up shown in figure A.5. Because of the variation of input impedance of the filter over the working frequency range, a buffer amplifier was used between the source and the filter.

The results are presented in figure A.6 and may be compared favourably with those from the CAD prediction.

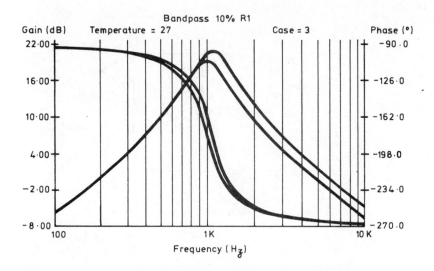

Bandpass 10% R1

| Gain (dB) | Temperature = 27 | Case = 3 | Phase (°) |

Frequency	=	100.00000 D+02 H$_z$		Gain	=	−4.925 dB
Phase angle	=	−267.678	Degrees	Group delay	=	994.43710 D−09 Sec
Gain slope	=	−613.66835 E−02 dB/OCT		Peak gain	=	21.023 dB/F
					=	104.5+01 H$_z$

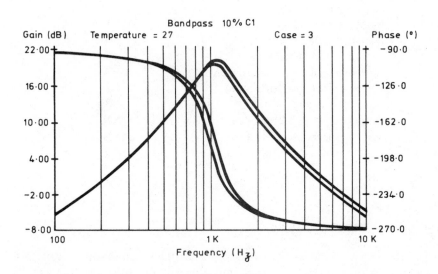

Bandpass 10% C1

| Gain (dB) | Temperature = 27 | Case = 3 | Phase (°) |

Frequency	=	100.00000 D+02 H$_z$		Gain	=	−6.718 dB
Phase angle	=	−268.992	Degress	Group delay	=	124.88288 D−88 Sec
Gain slope	=	−610.56437 E−02 dB/OCT		Peak gain	=	19.817 dB/F
					=	965.0+01 H$_z$

Figure A.3 Band-pass responses for changes in R_1 and C_1

0:Input resistance	Opamps Type 0. . .Alias Value	Tolerance (%)	Standard Components
0:Input resistance	1000000	0	Standard Components Library PDC
1:Open loop gain	2000000	50	0 :Opamps
2:Output resistance	75	0	1 :Diodes
3:Offset voltage (Voffset)	.005	50	2 :Bipolar transistors
4:Temp coeff. of Voffset (V/Deg C)	.000005	50	3 :MOS Transistors
5:First pole (HZ)	5	40	4 :Programmable waveforms
6:Second pole (HZ)	1000000	60	5 :Sinusoidal sources
7:Slew rate (V/Sec)	500000.2	30	6 :Transformers
8:Input offset current (Ioffset)	0	0	7 :Polynomial sources
9:Input bias current	0	0	8 :Printer copy of library
10:Current doubling interval (Deg. C)	0	0	9 :Passive component labels

Standard Components

10: Retrieve a library
11:Save library
12:Rename current library
13:Quit
14:Change colors

E:Edit   J:Jump   N:Next   L:Last   C:Copy   A:Alter alias   Q:Quit   Your choice?

Figure A.4 Op-amp specification and component library

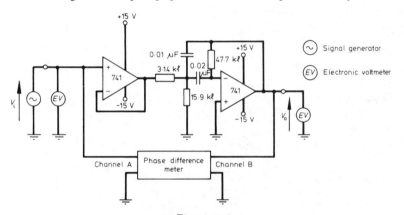

Figure A.5

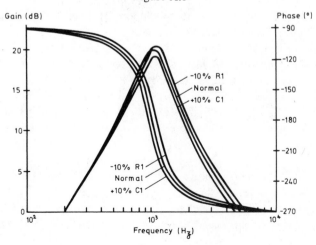

Figure A.6

Bibliography

Included is a short list for further reading. The first eleven entries are books and they are all excellent in their coverage of material. The book by A. B. Williams is very comprehensive and that by M. E. Van Valkenburg highly readable. The book by D. E. Johnson, J. R. Johnson and H.P. Moore contains charts and tables and the book by P. R. Geffe, although rather venerable, still contains good practical advice on passive circuits.

The remaining entries are technical papers, those by L. T. Bruton and A. Antoniou being mentioned in the main text. The paper by S. Darlington is of historical importance in connection with the synthesis of passive circuits.

1. Barna, A. and Porat, D. I. (1989). *Operational Amplifiers*, Wiley, New York.
2. Chen, C. (1982). *Active Filter Design*, Hayden, New Jersey.
3. Huelsman, L. P. and Allen, P. E. (1980). *Introduction to the Theory and Design of Active Filters*, McGraw-Hill, New York.
4. Van Valkenburg, M. E. (1982). *Analogue Filter Design*, Holt, Rinehart and Winston, New York.
5. Williams, A. B. (1981). *Electronic Filter Design Handbook*, McGraw-Hill, London.
6. Mitra, A. (Ed.) (1971). *Active Inductorless Filters*, IEEE Press.
7. Schaumann, R., Soderstrand, M. A. and Laker, K. R. (Eds) (1981). *Modern Active Filter Design*, IEEE Press.
8. Geffe, P. R. (1964). *Simplified Modern Filter Design*, Iliffe Books, London.
9. Johnson, D. E., Johnson, J. R. and Moore, H. P. (1980). *A Handbook of Active Filters*, Prentice-Hall, New York.
10. Bowron, P. and Stephenson, F. W. (1979). *Active Filters for Communications and Instrumentation*, McGraw-Hill, London.
11. Temes, G. C. and Mitra, S. K. (Eds) (1973). *Modern Filter Theory and Design*, Wiley, New York.
12. Darlington, S. (1939). 'Synthesis of reactance 4-poles which produce prescribed insertion loss characteristics', *Journal of Mathematical Physics*, pp. 257–353.
13. Bruton, L. T. (1969). 'Network transfer functions using the concept of frequency dependent negative resistance', *IEEE Trans. Circuit Theory*, **CT-16**, pp. 406–408.
14. Antoniou, A. (1971). 'Bandpass transformations and realisations using frequency dependent negative resistance elements', *IEEE Trans. Circuit Theory*, **CT-18**, pp. 297–299.

15. Temes, G. C., Orchard, H. J. and Jahanbegloo, M. (1978). 'Switched capacitor filter design using the bilinear z-transform', *IEEE Trans. Circuit Syst.*, **CAS-25**, pp. 1039–1044.
16. Fidler, J. K. and Nightingale, C. (1979). 'Slope normalised sensitivity – a new sensitivity measure', *Electronics Letters*, **15** (2).
17. Waters, A. and Newsome, J. P. (1985). 'The input impedance of voltage controlled voltage source active filters', *International Journal Elec. Eng. Educ.*, **22**, pp. 63–68 (Manchester University Press).
18. Schaumann, R. (1975). 'Low-sensitivity high frequency tunable active filters without external capacitors', *IEEE Trans. Circuits and Systems*, **CAS-22** (1).

Index